This publication describes parts and procedures to be used in the building of a Pro Touring car for off-highway usage. The information in this publication is for experienced and knowledgeable technicians only. The procedures described require specialized tools and skills. If you do not have the appropriate training and equipment to perform these procedures safely, this work should be performed by other professionals.

The information contained in this publication may change without notice. The publisher and Author reserve the right to make changes at any time in presentation or content.

The information contained in this publication is presented without warranty. All the risk for its use is assumed entirely by the user. The publisher disclaims all liability incurred in connection with the use of the information in the publication for elements beyond its control, like specific component design, mechanical procedure and the qualification of individual readers.

All trademarks, models and respective numbers, and other product designations shown are used solely for identification purposes.

ISBN: 0-9791152-4-8

Printed in the USA

Publisher info:
Whaba, Inc.
To order additional copies, go to sdparts.com or call Scoggin Dickey at (800) 456-0211

On the Cover:
My friend Larry Erickson, (designer of ZZTop's Cadzilla) did the cover design for this book. You can see samples of his work at www.externalcombustion.net
Thanks, Larry!

Photos are courtesy of GM Design, Mark Stielow and PowerPackNation.com

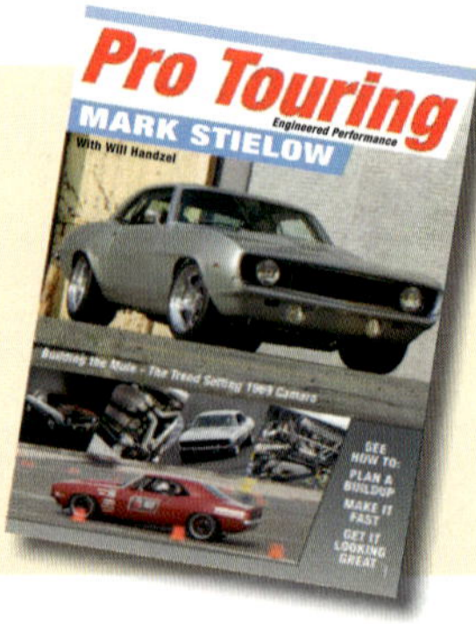

Table of Contents

Acknowledgments

I will say up front there are too many people to thank and I do not have a good enough memory to name all of you—so I will just say, "Thank You" to all that have helped me over the years. I have learned so much from so many great people and continue to learn more every day. Taking those learnings and creatively applying them in my buildups is one of the most satisfying aspects of my life and also what I consider to be the best way I can imagine to honor all those that have helped me. I'd also like to thank all of the companies that have supported my activities over the years with technical knowledge, parts and other help. Hopefully you have felt my thanks before through the product ideas I have shared with you, real-world driving experiences and my friendship.

I would like to specifically thank my parents, Don and Judy, for their unending support for my passion around cars. I would have never gotten to where I am without their help.

Also, I'd like to thank Will Handzel for pushing me all these years to document my activities. While I wince every time I have to stop what I'm doing to take a picture, the published stories lessen the pain a little every time I see them.

Finally, I'd like to thank Jack Chisenhall, Larry Erickson, Charley Lillard, Tom Baker, Kyle and Stacy Tucker, Ryan Kuhlenbeck, Dave Mikels, Larry Filipczak, Jim Fikse, Laura Howlett, Brian Baker, Joe Borschke, Whitey Brown, Brian Thomson, Ray and John Currie, Todd Gartshore, Gordon Rojewski, Joe and Paul Van Nus, Richard Schare and Cameron Evans.

If you've met me, you know I am not a very comfortable as a well-known person, but it is nice when I meet people and they share how reading my stories has helped them with their buildups. That's why Will and I did this book…with the hope it will provide just a little more detail on how I build winning Pro Touring cars for those looking for more answers as they step off into the unknown of building their own dream Pro Touring car.

Stielow

Here's where I have spent many nights…

Foreword

By Will Handzel

Back in the July 2001 issue of *Popular Hot Rodding (PHR)* magazine, I started the first story covering the buildup of a 1969 Camaro that would be called the 'Mule' with this:

'If there were a support group for thrashing car-builders, we would have to believe a meeting would go something like this. *"Hello. My name is Mark and I'm addicted to building wild '67,-8 & -9 Camaros, thrashing them hard and moving on to the next project." To which we'd all answer, "Hi Mark, you're in the right place." But there is one main difference between us and guys like Mark Stielow. Most of us will never build as many top performing and award gathering Camaros—hell, if we build just one car of the caliber he puts together, that would be plenty for a lifetime. No matter what you build, observing Stielow's thought- and build-processes will give you ideas to incorporate into your projects. Because of this, we are going to detail the building of Stielow's latest fast Camaro, a '69 vintage runner. He's named it the Mule since this car will be a testbed for the many ideas Stielow has had swimming in his head for the last few years. Look for more installments as the car goes together and some high performance testing when it's complete.'*

Reading that intro now years later, I am surprised how well I had Stielow's number! While Mark and I have been friends for about 15 years, I have never really thought much about us beyond our friendship and partnership as magazine story co-conspirators. Since the Mule debuted, its legacy as one of the most impressive project buildups *ever* has just continued to build: the car stands as a testament to Stielow's creativity, buildup discipline and fabrication skills and the 23 months of consecutive magazine technical stories in PHR are a record for longevity and continuum as far as I know.

In true fashion, Stielow didn't spend too much time soaking in the limelight of the Mule buildup. He has gone on to build other cars. He won multiple car shows and driving competitions with the Mule and has done the same with other cars he has built— like the LS9-powered monster called Jackass that he recently built for his friend Charley Lillard and the dominating 427ci LS9-powered Red Devil '69 Camaro he built last year.

While time and technology has marched on since the Mule Camaro was built, the build process is timeless…and immensely relevant. For this reason, Mark and I decided the hundreds of photos that never saw the light of day (we shot thousands of photos of that buildup), needed to be put to use to help all see the behind-the-scenes details of building a world class Pro Touring car.

So here it is, the book about the Mule buildup - the car that defiined Pro Touring.

Enjoy.

Will

Mark and I at the start of the Mule buildup

I'd Like to Thank the Following Companies for their Support Over the Years:

YEARONE

www.yearone.com
(800) 932-7663

THOMSON AUTOMOTIVE, INC. RACING ENGINES

www.thomsonengines.com
(800) 538-0044

CURRIE RACING REAREANDS

www.currieenterprises.com
(714) 528-6957

Auto Meter COMPETITION INSTRUMENTS

www.autometer.com
(866) 248-6356

RED LINE SYNTHETIC OIL

www.redlineoil.com
(800) 624-7958

The Buildup Plan

The Stielow system for completing a project on time and on target

To me, there is nothing better than spending time in your shop dreaming, creating and tuning on the next cool car... that's why I like my car building time so much...

So, you probably know I like to build early Camaros that I think look cool and perform to the extreme. Before we go any further, you should know that when I started out building the car shown in this book, the Mule, my goal was to build the nicest looking, best performing tri-year Camaro (that's 1967, 68 and 69) ever. The lessons I had learned building my other 5 tri-year Camaros and over 25 other types of cars before I started on the Mule showed me where I could improve the first generation Camaro and what major modifications to make to net real results in looks and performance. Whether you use this book to build an extreme performance car like the Mule, or just use various tech tips to improve the performance of your existing street machine, I feel my experiences can save you some money, improve the looks of your ride and make your car more fun to drive. Every car I've built has been for fun, and I definitely have fun driving them, so I hope you can have that kind of enjoyment with your car, too.

I do change the parts that will end up on a project if needed: like with this Kinsler injector system. It didn't initially go on the Mule engine but I went to it after this mock-up.

I do a lot of 'mockup' time when going through a project—actually seeing something in position on the car makes it easy to make a decision on whether to use it or not.

I believe a 'winning' Pro Touring car is measured in the details. To me, what makes a car appealing is not just what you see, but what you don't see. This is why I spend A LOT of time getting everything sitting 'right' on my cars--like the positioning of the wheels in the wheelwells.

Dreaming

Getting started with 'the dreaming' around a project is by far the best point, in my mind, of a buildup. You are not constrained by a budget or capability issues—you just dream about cool, amazing things and how fun it would be to have them become real.

The hard part goes from there: first you need to write those dreams down so you and others can understand them. Then, you need to figure out how you can manifest them to reality. Finally, you need to go out and make those dreams into a functioning, really cool, amazing vehicle.

This section is about the post-dream aspect of doing a project. I will leave the 'dreaming' part up to your imagination and give you some pointers on how to get those dreams documented to use as a guide with the project.

The following section will show you the forms and plans I, Mark Stielow, use to make a project happen.

Without a doubt, I am sure there are other ways to plan, track and complete a project. What is listed here is the system I have developed through the 30+ cars I have built and it has worked for me, so hopefully it will help you with your projects.

As a note, none of these documents will 'do the work' for you, or buy you the conviction needed to stay the course on a project when every 'friend' you have is pushing you to make changes to your original vision midway through (something I consider the main reason projects don't get completed or become abominations). These documents are merely tools to use. The work and conviction needed to complete a project on-time and on-target is something you either do or don't exhibit in your actions!

My Cars

To help you understand how I got to this point in my life, I present a sampling of the cars I have built over the years, starting with my high school cars.

I have owned a few Mustangs over the years. This '71 Mustang was my first car. I repainted it and found out it was slow—sold!

This '70 AMX was fun: a fast, simple, good looking, low mileage muscle car.

Actually won my high school drag competition my Junior year and got mentioned in the local paper with this photo: kinda fun, but the car was hit by a van on my way to High School.

Always liked the '67 Mustangs and this one was very straight. Had a 289ci small block and 4 speed transmission. The Centerline Cragar wheels and slapper bars were a requirement for the day…

My first play with a dedicated racecar started with this MG. It was a stripped Autocross car I got cheap. We Autocrossed it alot. I learned a lot about driving from that car. I traded it for a '70 Boss 302.

Vision

Once I am done dreaming up a vehicle, I write down a description of the car I've dreamed up. Then, I take and list the level of performance I expect it will produce on a vehicle. To do this, I use a document I call the Vision.

The Vision sheet lists all the things I want the vehicle to be able to do once it is on the road. It is the short sheet that I use as a decision guide when determining whether a modification/component change/system update/etc I am considering for the vehicle will help or hurt me achieve those performance targets.

As an example, if I have set a cornering goal of over 1G on the skidpad in my Vision, then the thought of running an 8-71 blower on the top of the engine would be a non-starter. The added weight of the blower and the location of it on top of the engine (and forward of the front axle) means my 1G skidpad goal is more difficult to achieve.

I know this is an extreme example, but actually, the Vision provides that kind of contrast throughout the build process. You just have to ask yourself if whatever you are doing will help or hurt you achieve your stated goals. Decision made.

It actually makes building the car much, much easier...

Ask yourself what are you going to do with this car? Then, the project just moves forward.

Vision

Vehicle Name : Mule 1969 Camaro
Date: 04/10/01

Dynamic Performance
0-60 mph Acceleration: 3.5 seconds
¼ mile ET/mph: 11.9 seconds/118 mph
Top Speed: 185 mph
Lateral Acceleration (skidpad): 1 g
Weight (with all fluids): 3150 lbs
Highway MPG: 19 mpg
Lap Times (at Gingerman): 2:35 min:sec
Braking from 60 mph: 115 feet

Appearance
- Look fast and be fast, while having a clean and purposeful appearance.
- Maintain 1969 Camaro appearance.
- Be built to last for a long, long time.

Engineering
- Incorporate a new suspension with less unsprung weight, more control and better tire contact patch maintenance than the stock suspension.
- Have the largest, widest wheels and tires possible. Fitting 275/35R18 front and 335/30R18 rear tires within 'stock appearing' wheelwells.
- Have all the creature comforts of a modern performance car: a/c, stereo, sound deadener & insulation & electronic fuel injection.

Completion Timing
-Needs to be completed in 24 months (or less) to be relevant!

Notes
I'll jump ahead and tell you the Mule accomplished these goals (and went beyond on a few), but it was a challenge.

I think the Mule is just now being recognized (10 years later) for the amazing leap forward in performance, integration and appearance that car manifested. While I have built a few Camaros since the Mule (X, Jackass, Red Devil) it will always hold a special place: it helped to move the Pro Touring crowd to build cars that 'worked' on both the street and the roadcourse/autocross track. This is something I had not considered, but am honored that it has that distinction.

My Cars (continued)

Oh, this Rustang was a real joy. Notice how badly swiss-cheesed the rear quarters were! I spent a lot of time doing sheetmetal work, sanding and painting this gem. Didn't keep it long...

But it looked pretty good after all that work, huh? I really like the look of these cars, but sold this one pretty soon after this shot.

My experience with other cars before I got this car allowed me to realize the first generation Camaros look, drive and work great. This was my first Camaro—an original RS car with 50,000 miles on it that I put a Doug Nash 5-speed in. What a fun car it was!

Planning

Long before I started on it, the Mule was something I could see in my mind...but it took a while for me to have the facilities, time or resources to build it. Being able to visualize a car before starting on it is critical to ending up with the car of your dreams. Too often enthusiasts embark on projects with infinitely vague goals in mind. Think about it, if you don't know where you're going, how in the hell do you expect to get there?

My suggestion is to write down everything you want your car to do when it's completed.

Conflicts like, "Have blown, injected Rat motor and corner like a modern Corvette" are clearly not going to work (the boat-anchor blower motor is like strapping an engine block to the hood). Come up with a cohesive set of needs and stick to them throughout the project. Hell, I suggest you come up with a name for the car as this forces you to picture it clearly: I show the format I use for this on the previous page.

I also recommend getting an illustration done to give you a truly visual goal or if you're on a budget, cut and paste some pictures to lock in the ride height, wheel and tire package and body shape. This also helps to keep you on track as you perform the buildup.

Build Shop

Once the picture is straight, the reality side of building a car needs to be dealt with before moving on. Having time, tools and space to work on The Mule was critical to me as I really wanted to build this car myself. By doing a lot of the work myself and extending my usually short build times (I have eventually sold all of my other cars to pay the bills from their construction), another goal was to minimize the money needed to complete the car.

Finances

This is an important point to discuss. Most car enthusiasts gloss over the financial investment required to complete the building of their dream car. To avoid a project being stillborn, before any work begins, the sobering task of doing a budget and timeline needs to be completed (I show examples of how I document my budgets and timelines on page 16 in this book). Often, the amount of money and time required to achieve the dream is beyond a person's means unless the project is spread out over a lifetime. I'm not trying to be a buzz-killer here—I want you to complete your car.

Avoiding financial ruin/divorce/homelessness is as important as having a cool car to drive to the cruise night, right? To that end, here's how I plan financially for building a car.

Make it Relevant

There is one more aspect to a buildup I consider key—relevance! If you take ten years building a car, don't be surprised if it looks retro by the time you complete it. This is why I work so hard to complete my cars in a short amount of time.

My Cars (continued)

Ohhhh, wish I still had this '66 Mustang fastback. What a sweet car. Had this one in college and drove it all over. Fun to drive and looked great.

This is another Camaro I had in college. It was powered by a ZZ3 350 ci. crate engine. I shifted through a Doug Nash 5-speed transmission and the car was equipped with a full Global West suspension.

Planning

Steps To Create A Project Car Financial Plan

1. **PARTS**—Make a list of all the parts you think the car will need. I use categories like Suspension, Engine, Transmission, Rearend, Interior, Body, Safety to further clarify where the parts will end up on the car. In the list, I include the prices, part numbers, manufacturers and phone numbers/web sites of where to get these parts. I put this all on an Excel file on the computer so it's easy to revise.

2. **LABOR**—Make a list of all the outside labor required to complete the car, like Sheetmetal work, Painting, Engine Building, etc. It's a good idea to estimate the time you will need to invest in various steps of the project, like Disassembly, Parts Cleaning, Reassembly, etc—if you are planning on doing work (this has a value).

3. **INVESTMENT**—From here, I total up all the money and time required and take it all in. Once you've regained consciousness, take a cold shower, go for a walk and come back to it a few days later.

4. **REVISE**—Keep working the "Vision" to come up with numbers that reflect reality. Continue to revise for a few weeks.

Once the numbers are based on reality, I suggest you build a timeline using your disposable income or willingness to go into debt, your available time/resources, the parts supplier's and skilled labor's ability to do the work required. Most of my cars take about 1000 hours of my time to build—and that might not be realistic for you as I know the early Camaro like the back of my hand. You should probably multiply your timeline by two or three to determine how many hours it would take to build a car from scratch.

Moneywise, you can spend anywhere from $10,000 to $150,000 or more to build a fun Camaro. One won't have paint and the other you probably would be afraid to drive, but both would still be Camaros.

To help you finish the project, I suggest you devise a timeline that keeps the project fun (not too long), but has milestones that are fun to revel in (mocking up the car for an 'in process' show appearance, firing the engine, preassembling the car before paint, etc). Nothing is better to spur you to complete more work than having 'the crew' over to check out the progress of your project, share in a few 'what ifs…" and crack open a cold one telling some stories. It's an opportunity to stop, enjoy the trip, then get back to work.

When I have settled on my money and time plan, I usually just take off on a project like a man possessed. Getting started is a struggle for me, so I like to get through the beginning phase as fast as I can. Once the car starts to take shape, I can calm down, but the beginnings are always tough. To do this, I clear the decks of all other commitments before jumping in (that means no coordinating of multiple projects at once!) and get the car to the point where it's sitting on all four tires looking like a car as quickly as possible. Everything from there seems downhill when you can look at the car and know it's a roller.

My Cars (continued)

When completed, the White Car had an 8 point rollcage, subframe connectors and late model Corvette brake parts. It was built to win overall in the One Lap of America race so it had lightweight, forged aluminum Fikse wheels, a 500 hp all aluminum small-block V8, ZF six speed and Recaro seats. I sold this car and it now lives somewhere in California.

I ran three One Lap competitions with this car and it finished a best of 4th overall in 1995—against some factory-supported programs. My goal was to win with an old car, which we didn't do, but we often ran up front, only to have something fail and knock us out (this car caught fire one year!).

Yes, that's me on the left and Kyle Tucker, co-owner of Detroit Speed & Engineering, on the right. We have been friends ever since we both attended college together in Rolla, Missouri (and both got jobs at GM).

The Parts & Labor List

I enjoy the process of finding the latest and greatest components that are available and figuring out how to integrate them into my buildups. These 'new' parts are just a fraction of the buildup, though. I usually do a large percentage of a buildup with components I have lots of experience with to maximize my success with the vehicle. The Mule was a departure from most of my previous builds as it was the car on which I decided to replace the entire suspension and drivetrain and modify just about everything else but the stock exterior.

Sample of Parts List

Description	Qty	Company	Price	Labor
Vehicle:				
Engine:				
Transmission				
Rearend				
Front Suspension				
Rear Suspension				
Brakes				

Sample of Labor Required

Description	Company	Cost	Hours
Disassembly:			
Sheetmetal:			
Etc.:			

My Cars (continued)

The engine in the White car was an all aluminum small block that really ran great. It was my first big foray into EFI and I loved it. Car had all the power you could want but would idle in traffic all day long.

QUADRADEUCE

A slight departure for me was the QuadraDeuce. I built this car when I was in the Engineering department at Summit. I really liked the idea of all-wheel drive in a roadster, so we gave this a try.

Here is the QuadraDeuce front suspension as it was going together: note the inboard coil-overs, independent front suspension with S10 Syclone front differential. These were early mockup parts.

The Budget

For good or bad reasons (depending on how you feel about spending everything you have on a car project) I have never let a buildup budget prevent me from building the car of my dreams. That's not to say I don't like to know what I'm getting myself into before I start, because I do. With that said, I can tell you that for just about every car I have built, my desires outstripped my ability to fund it. But I had an idea how much I would need from the start and put together a plan to raise the money or somehow traded-out my skills to fund the buildup.

To me, that's how you achieve your dreams. Reach as far as you can (or maybe a little beyond), but understand what the reality of the situation is before you completely step off the cliff. This is how we grow and achieve things we never thought possible.

For my budgets, I split the costs up into Vehicle, Parts purchases and Labor for services I need to have performed on the vehicle. In the Parts and Labor list, I make sure I have the Description of the component, who makes it, its part number, how much it retails for and how many I need, but in the Budget, I usually group these costs to simplify the chart.

The Budget chart does list when I think I will need each Part or piece of Labor in the buildup, as that helps me know when to make purchases or requests for labor early enough so that everything occurs as planned in the timeline—it also tells me when I am going to need money and how much!

My Cars (continued)

THE WHITE CAR

To save money, I towed this car back to Rolla, MO to have Neil Castle build the roll cage and subframe connectors.

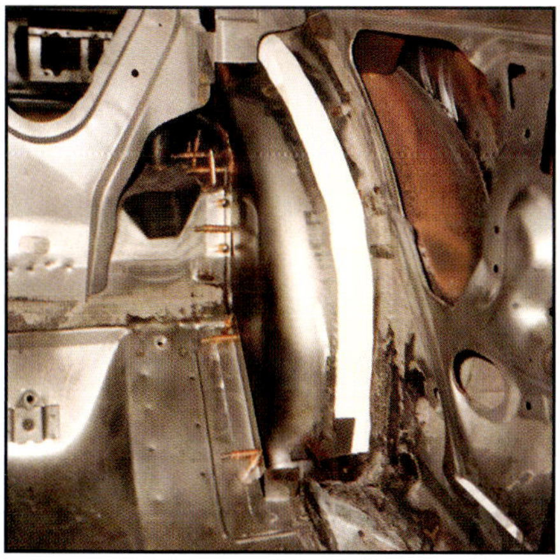

The Red Witch was the first car I mini-tubbed. Here is the wheelwell after it has been cut in half, and a paper template is being created to fill the gap. Hard to believe this is just standard practice now when building a Pro Touring car! Car ran ZR1 Corvette wheels and big BFG rubber.

The Red Witch got its name because it was a brutal buildup: I had made some timing commitments to HOT ROD Magazine and really killed myself to meet them.

I had my friend Larry Erickson do some drawings for the Red Witch: this was one of the wilder ones he did. I did delete the front bumper and run big rubber, but the changes required to get this ride height would have required too much work/money to achieve.

The Budget

Budget Summary Estimate

Vehicle (description):

Date:

Description	Amount ($)	Timing
Vehicle Cost	$____________	____________
Exterior	$____________	____________
Interior	$____________	____________
Drivetrain	$____________	____________
Suspension	$____________	____________
Brakes	$____________	____________
Steering	$____________	____________
Miscellaneous	$____________	____________

My Cars (continued)

The Red Witch was powered at first by a race-based splayed valve GTP engine during it's One Lap forays and then a Summit-built 532 ci, all aluminum big block on it's Power Tour runs–both engines were Accel fuel injected.

I ran this car in One Lap, but struggled with the splayed valve engine valvetrain. Power went through a ZF six speed, Currie 9 inch aluminum centersection rearend and it was all stopped by Baer brakes. Car had an ATL fuel cell poking out under the rear panel, a Harwood cowl hood and plenty of VW Red paint.

THE THRASHER

I moved to California for a while, but sold the Red Witch before I left the Midwest to a car guy via Hemmings. The beauty you see here was a low dollar, high desert hot rod. It needed lower quarters, but other than that was one of the most solid cars I had worked with up to that point. It drove fine, but that didn't stop me from chucking the engine, trans, rearend, suspension and interior for improved performance.

The Marketing Tradeout

There are so many bad rumors and misguided ideas out there around this segment of the business, I feel it is only right to tell how I 'promote' product.

First of all, for all the people that call these parts 'free', you obviously have never successfully completed a marketing tradeout of product…because if you had you'd know why I often feel like I'd have been better off if I just worked a second job and bought a part: I usually have so much time and effort in making a great technical install story, it feels like it just wasn't worth the effort of getting the part in the trade. I know in the end this is not true, but my feelings are a good indicator of how much work goes into creating a great technical story (or a great 200 page technical publication for that matter!).

So here is how I do a marketing tradeout: I meet with a company that has a product I would like to install during a future vehicle buildup and tell them I will use their product on my vehicle and show their product being installed on my vehicle in a magazine technical article (note: this is not the same as the car being shot for a 'feature' or included in a 'project car' story—those don't really help a company sell products: you need to do a technical story specifically on that product). In return, I ask the company to provide the part to the buildup: essentially trading the exposure of the technical story to the enthusiast market for the price of the product.

Many companies will not do this, some companies do, and the ones that do are very particular about who they work with and how this process works. There are tons of 'scam artists' out there that will get a company to provide a product to a buildup, but never really provide the exposure these companies need to sell enough product to 'pay back' the cost of providing the part. These folks are quickly exposed for who they are —the aftermarket automotive business is too small for those operators to hide in the shadows. Those characters usually only get away with this once and disappear from the business.

The hard part about doing a marketing tradeout is shooting the publication-quality photos, then putting everything together with words and other data to create an interesting magazine story. This sounds easy, but setting up all the shots and getting them lighted properly consumes lots of time and effort—and I have a lot of training from friends in the magazine business to show me what to do and not to do (a book in itself!). Then, for me, the writing part is just awful—like writing a term paper when all you want to do is go out and play with your car.

And all of this work isn't worth a nickel if you can't get a magazine or website to publish it—which means the story needs to be good enough they want it for their customers.

I am glad to have done the magazine articles over the years, as they have opened up a lot of doors for me and help dull the memory of the struggles encountered during every buildup, but they are still as tough to complete and get published today as they were 15 years ago…

My Cars (continued)

I didn't last long on the West Coast, but it was long enough to finish The Thrasher (this is the day it was completed, shot for a poster and put on a transporter back to the Midwest). I went with BBS wheels, BFG rubber (as on all my previous cars) and a four point cage. The little silver box hanging down under the license plate was the sump for the heavily modified, stock gas tank (detailed on page 63 of this book).

Here is the Thrasher as it was being shot for a magazine feature. This might be hard to believe, but this shot was taken in a parking lot early in the morning (the scrim is held up with some poles, and the plastic sheet laid out on the ground).

This car started with a very simple 400 ci efi small block V8, a T56 trans and a Currie rearend getting it to the wheels. Later, we swapped in a 18°-headed 427 ci. small block Chevy V8.

Project Starting Point

I have built cars using everything from badly rusted to incredibly clean Camaros and can definitely say it's worth the extra money to buy a solid car as the starting point of a project. Now, this is easier said then done, as I've bought what I thought were super clean cars, only to find rust in the bottom of the windshield frame, lower rear quarter panels and even the floors. Through hard lessons, here are my...

Shopping Tips for Buying a Rust-Free Car:

1. Shop in arid areas of the Southwest

2. Be brutally honest with yourself about whether a car is solid or not and take a pessimistic friend with an icepick and a magnet if you really want to know how solid something is.

3. Look for cars that have been painted two times or less, a goal that is harder than you think to achieve!

4. Avoid the swap meets and car shows, buy from single parties off Craigslist, Autotrader or out of the Recycler papers as there is less chance of someone turning a car for profit (can you say Earl Scheib?)

5. Don't worry about what drivetrain it has, the suspension condition, paint combination, or other easily replaceable items. A rust-free, straight car is the main goal.

I think of the body like the foundation of a house—it has to be solid and straight or everything I do to it will be compromised. My feelings are simple, this is a vehicle I'm going to spend more than money on, I am going to give it a piece of my life and it better be worthy of my time.

To that end, the Mule was based on a '69 Camaro I bought out of Southern California. It spent the last decade and a half of its life as an unfinished project sitting in a backyard. The body had some rust in the rear trunk extensions, front fenders and the right rear fender, but the floor and firewall area were absolutely as solid as the day it rolled off the assembly plant floor, which sold me on it. Having a solid car in the floor and firewall area is critical to achieving a tight, rattle-free car that won't rust in the future. To fix the rust in the other areas, I ordered replacement sheetmetal from CARS, Inc, in Rochester Hills, Michigan and had it installed at Reiter Metal Craft, in Mount Clemens, Michigan, but that is skipping forward in the story a little.

As I stated earlier, the purpose of this book is to give you a useful guide to building any kind of a one-off performance car by showing how I built a premiere Pro Touring car: the Mule 1969 Camaro. I enjoyed building the Mule and putting together this book. Was either one easy? No. But they are both work the effort.

I hope you enjoy reading this book and learn from it. If I can help you save time or money, minimize frustration and/or make your car faster and look cooler, then we both have won out.

My Cars (continued)

This is what a lot of my project cars look like in the beginning: I pay a little extra to get rust-free cars that are largely unmolested. This '69 Camaro was actually the starting point for Camaro X (see next page for what it looked like finished). I start by stripping the car completely down to pieces the moment I get it home.

Car being dipped

Not everything I do works out perfectly. I found a company that would ecoat a hot rod and introduced them to the world in the Mule stories. Unfortunately, their promises and performance did not match, so the company did not survive.

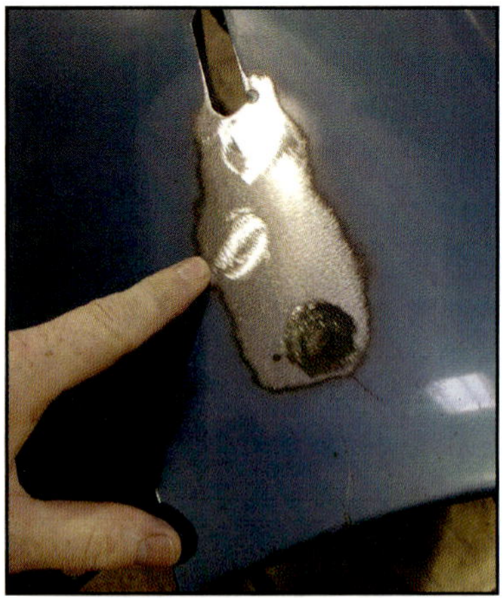

Metalwork

I like to have all of the sheetmetal modifications and repairs done at the beginning of the buildup. In this case, the work was done by the pros at Reiters Metal Works in Mt. Clemens, Michigan.

My Latest Cars

The tenth car I built was called Camaro X. It was the first '69 I built with LS power--a 427 ci LS7. The engine is light, makes excellent power and gets over 22 mpg on the highway. Incredible! I sold this car to Charley Lillard and he painted it blue.

I left the factory 'dry sump' oiling system on the Camaro X LS7 engine, which required some changes to the inner fenderwells to position the oil hoses for this near stock 550 hp LS7. This package made me a complete LS V8 convert. I got this engine from SDPC and Thomson Automotive tuned it up to 550+hp for me. Once I drove this car, I decided it would be LS-power from then on--they are impressive!

The front suspension underneath Camaro X consisted of a modified factory subframe, QA1 coilover shocks, DSE A-arms, American Touring Specialties (ATS) spindles & Baer brakes. I did some consulting work with ATS on their spindles and they are still available today--this system worked amazingly well.

Jackass
My friend, Charley Lillard, wanted his own Pro Touring Camaro, so I built this car, what he named 'Jackass', for him. It was my first 'customer car'...and probably my last. Nothing against Charley, but I like to build cars for myself and found that out by doing this car. The car turned out great and hauls.

Red Devil
My latest car looks familiar, but is different in a few key areas. First, it is powered by an 760 hp 427 ci LS9 V8 that idles at 650 rpm and looks like stock! The vehicle is also equipped with the first (for me) Pro Touring anti-lock braking system (ABS). The front and rear suspension are from Detroit Speed. This car is built to go incredibly fast and make it all look easy! I won the LS Fest & Optima Shootout competitions with this car - it is quick, fast and fun to drive!

The Shop

The tools, parts and garage I use to build my Pro Touring cars

What you see here on the next few pages is the nicest build area I have ever had to design, engineer and construct my cars: a separate 3 car garage in my backyard. I will tell you, the majority of my cars were not built in this location: for those, I bounced around in other people's shops or did the work in the unheated, poorly lit garages that were attached to my house. So, what I'm saying is a shop like this is not a requirement to build a winning Pro Touring car...but it definitely helps!

I built this garage after saving up a bunch of money for it. To me, if you are only going to build one project car in your lifetime, a full-blown out-garage like this is not needed. But, if you plan on doing a few projects, this is the way to go. It is a real luxury to be able to walk out of my house at any time and make real progress on a project for an hour or so—and I always am thankful to have it so close and so well stocked. There are only a few things I can't do in this garage (like paint a car), but those are things I don't like to do anyway.

While I understand getting a dedicated shop is a big one-time commitment, the acquisition of tools and support materials can be done over the course of the years for small amounts of money and sweat equity. To give you a little background, before I had my shop, I had already acquired just about all the tools and parts you see here.

I recommend you start gathering tools and parts now to support your builds. That way, if you are someday lucky enough to have a shop like this, you will have all the resources you need to fill it.

You'd be surprised what you can build in a 3 car garage! All the fabrication and finish work for the Mule was completed in this garage (the bodywork was done offsite). Here is a walkthrough of all the things in my garage and why I think they help build a great Pro Touring car.

Shop Floor Heat

There was one thing I got for this garage that some might consider a luxury, but I don't: floor heat. Before the concrete floor was poured, we laid out a grid of hoses that I now recirculate warm water through in the winter to heat the floor. I have a separate heater for the air, but having the floor at about 45 degrees makes all the difference in the world! This makes the shop incredibly comfortable to work in when it is cold as hell outside (which it gets where I live). I would recommend this to anyone building a shop.

Utility Sink

This seems like a small thing, but I like being able to keep my hands clean without having to constantly tromp back into the house. I use a small 2 gallon electric water heater so I have hot and cold water.

Combo Belt/Disc Grinder

I bought this grinder years ago at a used tool shop and have used it during the fabrication of literally thousands of components. It generates a mess, but allows parts to be smoothed/rounded/trimmed/etc superquick. I couldn't live without this now.

Band saw

I have had this band saw for years and it is another piece of equipment I don't know how you could live without. Safety needs to be a priority with one of these as they cut everything that gets to the blade: always use a piece of wood or metal as a pusher.

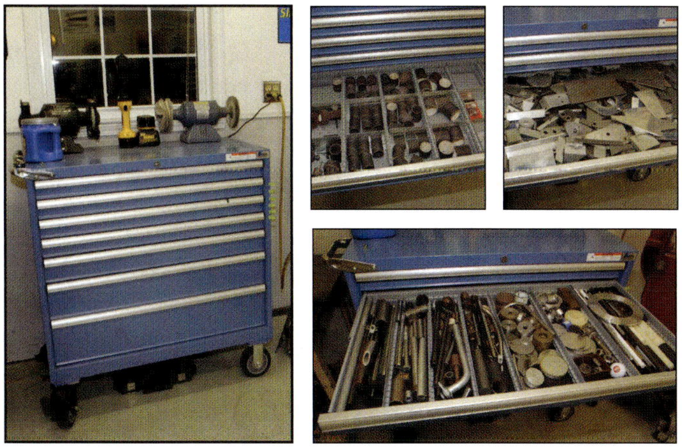

Fabrication Cabinet

I like to have plenty of supplies and raw materials on hand, but I don't like to look at or have to search for them. I bought this cabinet used and have filled it with all my smaller pieces of raw metal plate and bar and the metal finishing equipment (grinders, grinding discs, etc).

Hero Wall

This might not seem like it is important, but it has helped me over the years. When I first started out building cars, I would put up posters or photos of cars I saw in the magazines. Then, after I got my cars in the magazines, I would put up the pictures of those. Looking at these gets me through the frustrating, difficult times that ALWAYS occur during buildups.

Wire Wheel, Grinder, Buffers

I will occasionally build something and wish it looked more finished, so I will shine up the surface with these wire wheels and buffers. I also keep my rechargeable shop light and recharger here as it is a very easy location: I know to put it back there so when I need it, I can quickly find it.

Vice Bench Table with Clamps

This bench is a $100 item from an auction and something I use constantly during a buildup. I like it because it is heavy, is on wheels, has two huge bench vices welded to it and has a thick metal top. It doesn't look nice, but part of that is because I do a lot of work on it. I have been acquiring hand clamps (hanging on the wall) for years.

TIG welder

You don't need a massive water-cooled tip TIG welder like what I have but they allow you to build anything—as long as you know how to use it. I have one because I like to build critical components (suspension, rollcages, etc) that need a lot of heat penetration in the welds to be strong. Otherwise, a small MIG welder will work fine for brackets and sheetmetal work.

TIG Welding Rod

I keep a full assortment of welding rod and welding tips on hand for aluminum, steel and stainless steel. This seems like overkill, but I think one of the big project 'killers' is having to constantly run out to buy parts, tools and consumables one at a time instead of just having what you need on hand.

Surface Plate

My surface plate is a funny thing. I bought this at an auction years ago, then sold it because I thought it took up too much space (which it does)…but I missed it, so I bought it back (which is saying something because it weighs over 1000 lbs!). It is great for mocking up components 'in position', which I do a lot.

Spare Parts

I am a scrounger and tinkerer, so having spare parts laying around allows me to cut stuff up and try things without it costing a fortune. Most of my parts are stored under my surface plate table and in the attic of my garage (which was built to withstand the weight of having parts up there) in large plastic bins like these.

Bridgeport Mill and Machining Tools

I felt like I had 'made it' when my Bridgeport was delivered. Like the TIG welder, it is overkill for most people, but it allows me to make just about anything I can dream up. I didn't have many tools for it in the beginning—but I have now filled my old Craftsman toolbox with tooling. Before my mill, I had a Craftsman benchtop drill press that I about wore out—I definitely got my money out of that!

Clausing Lathe

I like this because I can make bushings, turn down a part that is not fitting (like a collar or spacer) and, in general, build anything between the mill and this lathe. I bought this used and also like it because it looks similar to the mill I have. Both the mill and lathe have digital readouts on them and are very accurate.

Shop Vacuum

I am a neatnik in my shop. It rarely looks like there is work going on because I am constantly cleaning. To make that easier, I bought this large, two-motor vacuum years ago and love it because it sucks up everything in its path. It is always plugged in with a 50 foot cord and a long hose, so I can go anywhere in the shop with it…

Toolbox

There's nothing better than having the right tool for the job, so I have just about every tool you can imagine and, to hold them, a massive toolbox. I believe in buying quality tools, so I now own almost entirely Snap-On tools (though I started like most with a complete Craftsman set). If you are going to work on cars, get quality tools.

Fitting and Fastener Cabinets

I also have a bunch of AN and brake fittings, which I now put in the gray cabinet I store on the top of my toolbox. I have purchased many AN fittings over the years, but didn't like using them because they weren't black—so I recently had all of them stripped and Anodized black! Problem solved.

Fastener and Fitting Assortment

Over the years, I have purchased a large assortment of fasteners for my buildups. I used to store these in totes, but it got to the point that I was accessing them so much, I bought a large cabinet to put the fasteners.

Stock & Factory Fasteners
I have saved every useable bolt, nut and clip off every car I have ever worked on. All of them get chemically stripped and coated before being separated and stored in drawers like these. This way, I always have the right fastener for the job!

Cabinet
For safety and organization, I bought a cabinet and that is where I store any brake clean, spray cans or other petroleum-based products. Anytime you are welding or grinding around petroleum products, you can't be too careful!

Wiring Supplies
Also, I have a full shelving unit with wire, connectors and other wiring support materials—you can't have enough wiring materials on hand if you are going to be wiring a vehicle that has any electronic-controls.

One way to force myself to take the photos needed for stories is to always have a 35 mm-style digital camera sitting ready on a tripod in the shop. In the beginning, stopping to shoot photos was a pain, but over time, it has gotten to be second nature. I suggest you do this, as it is cool to go back and look over what you were doing later on.

These ceiling mounted power and compressed air cord retractable reels were added just a few years ago, and I am not sure how I lived without them. They keep the cords off the floor and make it easy to get power and compressed air when and where you need them.

The Attic Storage Room
My shop has an upper area that was built to hold parts, my floor heat system and the air compressor. I use a retractable attic ladder to access it and have wired lights into the area to be able to easily see what is up there. By the way, my opinion is that you need an air compressor—how big is up to you, but you need an air compressor. If you are doing paintwork, you'll need a big one that pushes a lot of air. If you are just using it to power hand tools, you can get by with one like this. This compressor is at least 20 years old and works great for my needs.

The Engine Hoist

I have an engine hoist that I store in my attached garage that I bought 15 years ago. It is one of those required tools if you're building a car that takes up a lot of space, but also pays for itself in what it can do. It has more than paid for itself.

Stuff Bins

Ever since I started messing with cars, I have kept an assortment of fasteners and parts on hand 'just in case'. And you know what? Even today, I find it very handy to have a bunch of bolts, brackets and other stuff on hand to mock something up and help me 'think' through an idea I have. Usually, it is stuff I pulled off a project vehicle or gathered at the local 'Pick Your Part' salvage yard, so I have just a few bucks in it. This makes it much easier to cut stuff up and schmush it together to test an idea.

I can't even remember where these bins came from, but I love 'em. I fill them with stuff, tape labels to them and store them under my surface plate, so they are out of the way.

This bin is filled with brackets and pumps I use to mock up front drives on engines once I get them installed in a car. I like the bins with the pivoting, two-piece lids. You can buy these at home improvement stores, auto parts stores and other stores for a few bucks each.

The ubiquitous BOLT BIN!

This is my old style bolt bin—nothing classy, just a bunch of stuff I have gathered up over the years. When I have time, I go through and sort these, send them out for stripping and plating, then place them in separate bins in drawers

Here's what I have in separate Stuff Bins:

- *Factory Bolts/Nuts/Washers/Clips/Etc*
- *Factory Accessory Drive Brackets/Pumps/Etc*
- *Fuel system components*
- *Electrical components and wire*
- *Interior fasteners*
- *Brake components*
- *Various aftermarket components*
- *Sheetmetal (Aluminum, Mild Steel, Stainless)*
- *Plate (Aluminum & Mild Steel of thicknesses from 1/8 inch to 1 inch)*
- *Plastic sheet (black and clear of various thicknesses and properties)*
- *Tubing (Mild Steel & Aluminum of various diameters and thicknesses)*

Custom Suspension Buildup

Building a Pro Touring rear suspension from scratch

From my first thoughts of the Mule, it was a car with a custom rear suspension in it. Every one of my previous Camaros had used a version of the factory leaf spring/solid axle rear suspension, but I was not satisfied with the ride and handling of those cars. I had done everything in my power to make that suspension design work and decided I wasn't going to compromise again.

The problem with that decision was that in 2001 no one offered a rear suspension built for all-around performance (there were drag racing rear suspensions, but those were only good for going straight).

So, I built one. And I think I did a good job because it worked great and there are rear suspensions you can now buy for tri-year Camaros that look a lot like this one! This section will explain how I built this suspension and why I did what I did. Also, this shows how I take a design idea to reality—a process you can use if you choose to build an entirely new system for your vehicle (whether it be suspension, drivetrain, interior, etc).

Here is the Mule rear suspension I built in 2001: long before you could buy a four-link, coil-over, all-around performance suspension for a 1967/68/69 Camaro. The aluminum centersection Currie 9-inch axle used here has been in a bunch of my cars because it is bulletproof, looks good, is reasonably priced and is lightweight.

◀ *The basic design of the Mule rear suspension was to use the front leaf spring mounts (arrows) for the lower link body mounts, weld 'load boxes' into the floor for the upper link body mountss and weld a U-channel 'load box' above the rearend for the shock absorbers and panhard bar to mount in. It has been durable with 1000+ hp spinning 335 mm wide tires, so I think it works!*

This is the rear of the Mule before I started cutting on it. The rear wheelwells have been widened (much easier to do now with Deep Tubs from Detroit Speed & Engineering) but otherwise, this is the stock underbody. Do not take cutting up a perfectly good car lightly! This is scary stuff.

To make it easy to locate the mounting points, I used the dimensions from the suspension design I created in my computer to build this simple fixture out of ¾ inch diameter round, $^1/_8$ inch thick mild steel tubing. This is what I used to mock the entire package up in the car.

When I'm building a car, the best thing for me is to have something I can touch and feel: thus my mockup rearend built from a straight length of 3 inch exhaust tubing. I welded some sheetmetal plates to it to replicate the four link mounting locations and bolted it in place on the car via the snubber holes on the framerail (arrow).

Using the suspension link fixture as a guide, I cut holes in the floorplan to allow the upper links to protrude through and then I went to work fabbing 'load boxes' out of 1/8 inch thick mild steel plate. These were welded to the floor/subframe structure to mount the links to the body.

The cool thing about using my mockup rearend is I can quickly test multiple combos of mounting points, ride height, clearance and other aspects with a simple 'cut/weld/bolt-on' process—minimizing the loss of time or money and every car builders friend: frustration.

I actually built this mockup rearend before the Mule buildup to determine the axle width for specific wheel/tire combos. It has a set of junker axles that slide into a small hole on each end of the tube (finger) and hose clamps on the axles to set the overall axle width

With my mockup rearend, I tested various rearend widths long before I ordered the Currie 9-inch. Here I've got the mockup rearend installed and am setting a test width on it with a tape measure and adjusting how much the axles protrude by adjusting the hose clamps on them.

I always test the rearend width with the disc brake rotor installed (in this case a Baer Brakes 13 inch rotor) on the axle flange. The wheel/tire package here is a lightweight forged aluminum 18 x 12.5 inch Fikse wheel with 335/35R18 Michelin Pilot Sports.

With everything bolted on, I measure how the wheel/tire combo sits in the wheelwell. I like to position the wheel/tire combo so the tire is ½ inch from the lip of the outer sheetmetal and it is ¾ inch away from the inner wheelwell sheetmetal at the vehicle ride height.

One of my big issues is when cars don't have the wheels/tires located in the center of the wheelwell opening, so I measure the front and rear gap (shown) to make sure they will be centered. With this settled, I ordered the rearend from Currie and continue working.

With everything located, I was able to fabricate and finish weld the load boxes to the floor of the Mule. These look like they prevent the use of the stock rear seat, but I modified the seat frame to fit over these load boxes—looked stock!

Bracket Building - *Skipping forward a few weeks, the Currie rearend has arrived and I am fabricating and welding the brackets on the tubes of the rearend housing. The bracket being shown is the lower mount for the QA1 coilover shock absorbers. I added multiple holes in the brackets to allow for ride height adjustment.*

Bracket Building

Starting with a design I created on paper, I cut a piece of 3 inch wide, 0.125 inch thick mild steel plate (purchased to make into mounting plates) a little longer than needed and marked off the center of the holes I wanted in it.

I like to use my Bridgeport for this type of drilling as it has a digital readout on it so I can position the holes exactly where I want them. Notice the 3 inch hole saw on the table (arrow) I used to cut one end of the bracket.

To remove any sharp edges, I pull the sharp edge of a fabricators deburring tool around the inside of the holes. Notice the holes aren't centered on the bracket: this is to offset the centerline of the coilover shock so it clears the rearend housing.

To be able to transfer my designs onto the metal part with a scribe (the only way to get accuracy), I spray it with blue layout fluid and let it dry a few minutes.

I then locate the centerline of the part by measuring 1.5 inches on my calipers and dragging one of the caliper legs down the part edge, the other dragging a line down the center of the part.

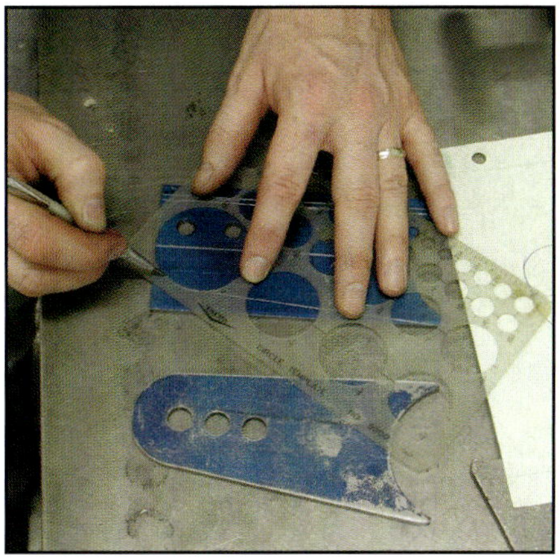

Using a suspension link bracket I built as a template, I copy the trim line and radius onto the new bracket using my mechanical drawing templates and rulers from high school and a scribe.

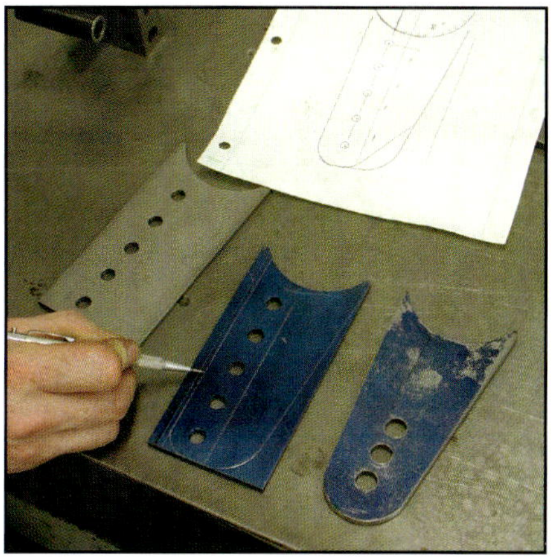

Here is a good representation of my progress: on the left is the raw 3 inch wide plate with the holes drilled in it, in the middle is the same bracket but with the scribe marks on it and on the right is a finished suspension link mounting plate. Notice the hand drawn design, you don't need a trick computer program to draw something up!

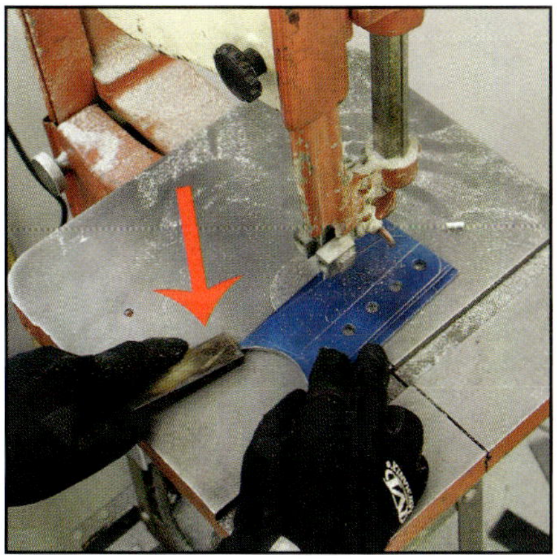

I then fire up my trusty bandsaw and cut away the metal that doesn't look like a bracket. Notice the 'pusher' piece of wood (arrow) used to keep my hand away from the blade—always use one!

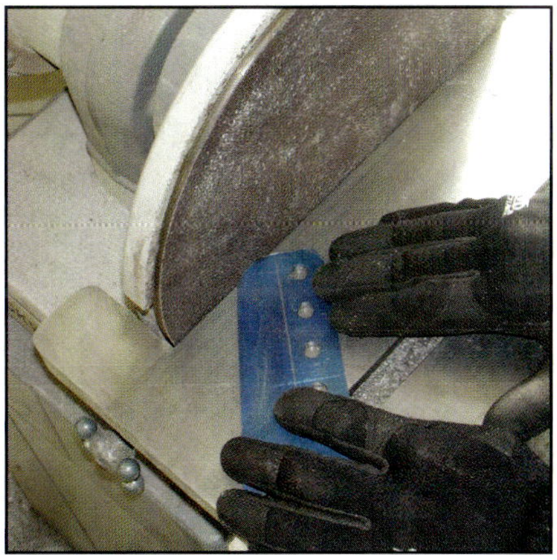

This is where a floor sander comes in very handy: I then take the bracket over to the belt/disc sander and lightly tune up all the edges so they are smooth and free of burrs. I am not trying to remove material here, just make everything look good.

I then test fit the brackets on the rearend with the suspension link fixture shimmed in 'car position' (notice the machining block with the small shim under it—see arrow). Any fitment tuneups required will be addressed now with the grinder.

Here is another reason I love my surface plate. If I want to transfer a dimension from one side of a car to the other side, I simply measure off the surface plate with this measuring stand, then flip the part over and scribe a line on the part. Done!

To make this line, I rotated the housing around while the edge of the measuring stand was touching it. Notice the housing end is only tack welded in place: I final weld everything in place on the rearend with it fixtured in place to minimize warping.

Here's another great tool if you are going to be fabricating anything: machining blocks. These highly accurate blocks are machined to specific sizes and can be stacked and shimmed to provide exact heights on parts sitting on your surface plate or mill. I bought all of these used and put them to work all the time!

So here is the Currie rearend housing coilover shock mount test-bolted on the shock bracket (shown here behind the lower suspension link bracket). Notice I have sculpted another section out of the bracket to match the shape of the QA1 bracket (arrow).

With everything test fit in place, I tack weld the brackets in place (with my TIG welder), pull the various components out of the way and remeasure everything to make sure it is all in the location I think it is (and want it to be).

I then begin the lengthy TIG welding process: weld one end, move to the other end of the housing and weld a similar amount, let housing cool, repeat until brackets are welded in place. Notice the lightening hole (arrow) I have drilled in the lower suspension mount.

An Easier Way

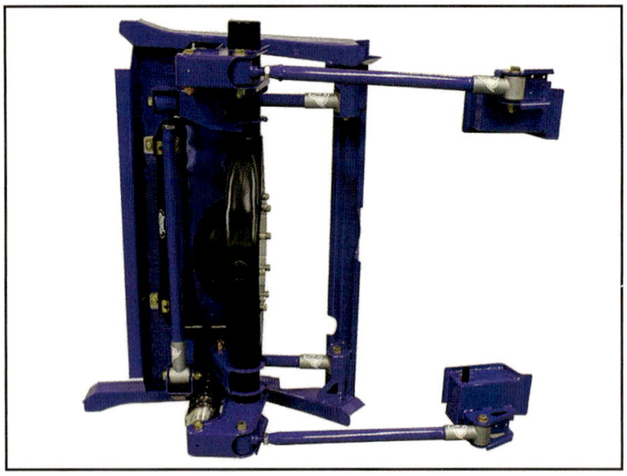

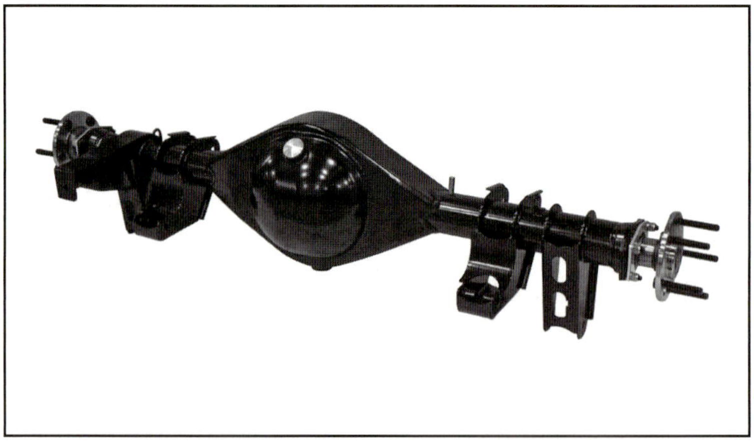

Luckily, if you are building a 1st or 2nd Gen Camaro,or 1st or 2nd Gen Nova into a Pro Touring car, you won't need to build a rear suspension for it like I did. You can just buy the suspension and mini-tubs from Detroit Speed & Engineering (DSE). My good friend from college, Kyle Tucker, and his wife Stacy, have done a great job building DSE into an excellent source for high quality Pro Touring components and systems. DSE offers a rear suspension for tri-year Camaros that is a great choice whether you are going to drive on the street or on the racetrack. You can find them at www.detroitspeed.com.

Building a Suspension Link

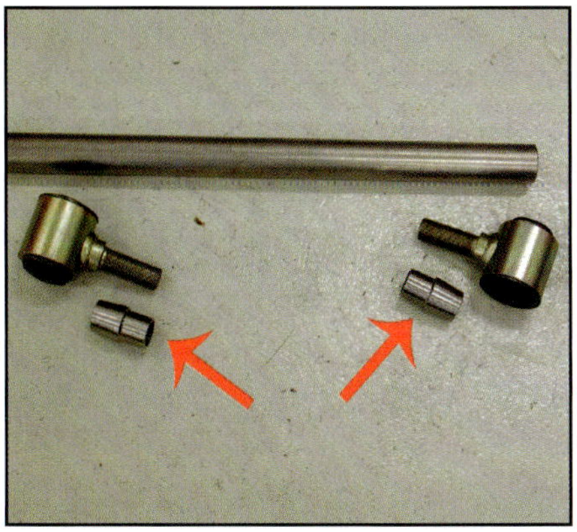

To start building this suspension link, I choose 1 inch diameter, 0.065 inch wall mild steel tubing to connect to the thread-on 5/8 inch Coleman Engineering end bushings that connect to the rearend housing and body structure. The 0.065 inch wall tubing saves weight, but requires the bushings.

I measure out the total length required for the link, then subtract the 'stackup' length of the adapter and thread-in link to determine the required length of the tube. I measure the tube and mark it with a Sharpee pen in prep for cutting it to length.

To cut the tubing, I use a 'chop saw' I bought about 20 years ago by locking the suspension link tube in place on the saw (putting the line on the inside edge of the saw to end up with the proper length on the long end of the tube) and cutting it.

The chop saw is not a precision tool, but it does a great job at quickly cutting tubing. The end will need to be tuned up on the grinder: I do this by laying the tube on the disc table and flattening the end by sanding it flat. Then I tilt the tube up (shown) and, while rotating the tube, grind it to bevel the outer edge.

I remove any remaining cutting remnants of metal still on the tube by pulling a fabricators knife around the inside diameter of tube.

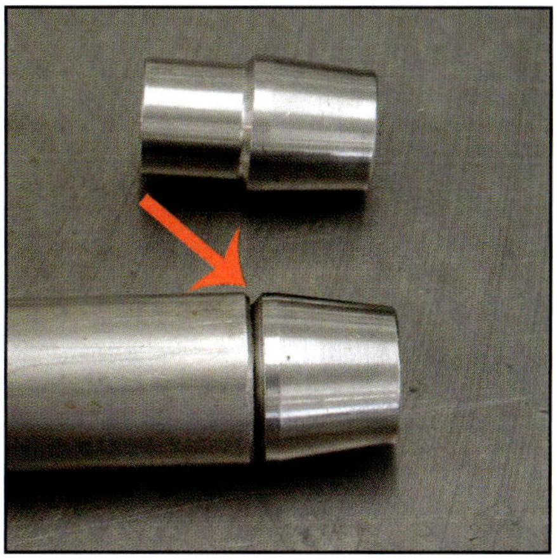

I purchased these threaded tube ends from The Chassis Shop. These are made to slip into the tubes. Notice the tube and adapter are beveled where they mate (arrow) so I can lay a thick weld bead down to join them together.

To check my work, I then position the suspension link on my link fixture. Notice how I have 'jamb nuts' (arrow on one) on each end to lock the link length and that there are an equal amount of threads showing at each of the link. I then built the other three links in the same manner.

I also machined up these spacers out of ¾ inch mild steel bar stock to fill the opening in the front leaf spring mounting bracket (which I used for the body mount for the lower suspension link). The mounting bolt threads through the center of this.

Building the Shock Absorber and Panhard Bar Mount

With the suspension link mounts created on the vehicle and rearend and all of the suspension links built, I turned my attention to creating a place to mount the coilover shock absorbers and panhard bar (the link that keeps the rearend in position from side-to-side). My solution was to have a 48 x 12 inch wide, 0.125 inch thick mild steel plate bent into a 'U' shape to make a 'load box' that I could then weld in the car. This section shows how I did this.

Here I am mocking up my 'load box' for the coilover shocks and panhard bar. This load box will need to hold the weight of the vehicle, so I am going to carefully weld it to the framerails, wheelwells and floor to tie it firmly to the structure of the body.

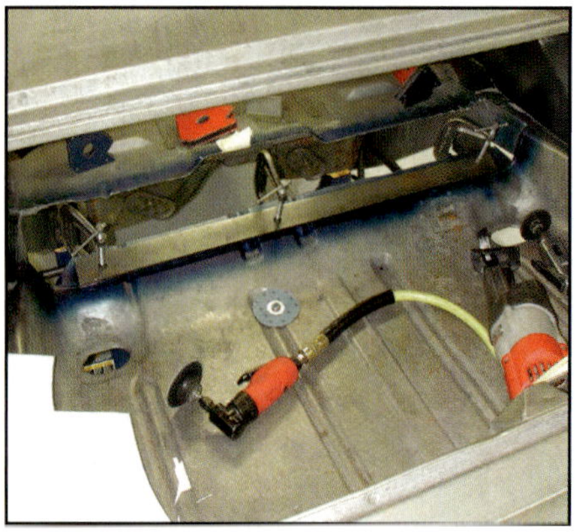

I started trimming an opening for the load box with my trusty sawzall, then used a small cutting blade to trim the tight spots. I clamped a plate to the trunk floor to stop it from flopping around when I was grinding on it.

To make sure the opening was exactly to the size of the load box, I used a carpenters level positioned on the back of the tube of the mockup rearend. To do this, you need to have the car level front to back and sided to side (I set the lower sills level on the side of the car.

It is nearly impossible to 'add' sheetmetal back to the car, so when the opening size starts to get close to the proper size, I do a lot of test fitting. In this case, that means maneuvering the load box onto the opening to see where it needs to be trimmed more, then pulling the load box back out and continuing the trimming.

In the final stages of creating the opening, I use a hand file to make it hard to take too much metal off. If this sounds like work, it is, but it's worth it if everything goes together tightly. My final 'fitment check' is when the load box fits in the opening with no gaps around the metal.

One advantage of having a 220V TIG welder and knowing how to use it means this load box could be welded up in my shop. As an FYI, when welding suspension components, my experience has been that a 220V welder is needed to get enough weld penetration.

To mate tightly to the wheelwells and frame section on the ends of the load box, I trimmed the U-channel and added a closeout panel (being welded into position). Also, notice the double shear mount welded in place for the coilover shock (arrow).

Here is the load box channel final welded with the panhard bar mount installed (arrow). This is now ready to be loaded into the vehicle and welded in place.

I positioned the load box 'level' in the car and applied a few tack welds to hold it in place. Then, I alternated my welding on each side of the 'box to minimize warping.

Setting Shock Length

The QA1 coilover shock absorbers require some assembly when they arrive. I always wipe grease on the threads of the shock body before threading everything together.

I also built a panhard bar and bolted it onto the body mount. Since I didn't have a panhard mount on my mockup rearend, I used safety wire to hold the bar level on the car to get my measurements for the Currie rearend. Then I built the bracket on the rearend.

With all the mounts final welded into the car, I asked my friend Larry Filipczak to help me load the rearend suspension in to the car (friends helping out with a build is one thing that makes car building fun) with mockup fasteners so I can see what it looks like put together.

I then used a floor jack to run the suspension through its full travel a few times to see how everything worked. Notice the panhard bar mount (arrow) on the rearend housing: it doubles as one side of the lower link mount (and how heavily drilled it is for lightness: details!).

I have ziptied the suspension in position so it hangs at ride height. This way I can look over my work. In general, this is the system I ended up using for the Mule.

I put a few adjustment holes in the upper and lower brackets just in case my original theory about the suspension didn't work out. Luckily, I never had to change these as the car handled amazingly well from the first few miles we put on it.

This angle gives you a good idea how tightly I positioned the upper suspension link to the inside edge of the body structure framerail (arrow) on the Camaro. This way, I could tie the load box to the framerail for maximum strength, while placing the link as far outboard as possible for maximum suspension control.

A panhard bar provides the optimal suspension performance when it is long (check!) and when it sits 'flat' when the vehicle is at ride height (check!). The long bar and flat positioning minimizes the change in location for the rearend as it moves up and down.

TIP: *Now that I know everything fits properly, I note my fastener requirements (so I can buy Grade 8 fasteners) and remove the parts so they can be prepped and powdercoated.*

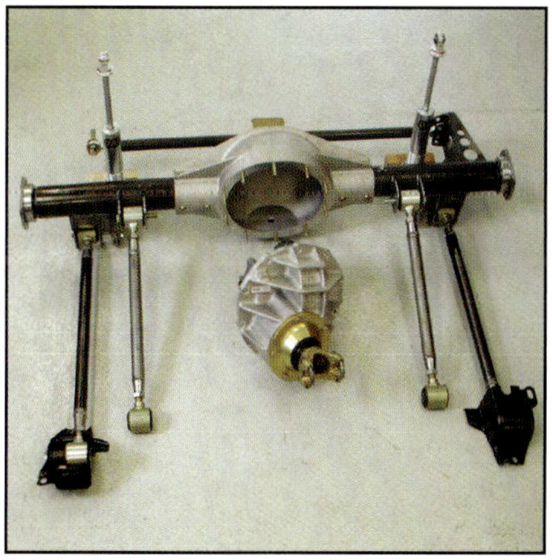

Here is the Mule rear suspension before it went out for powdercoating. I am proud of how it went together and how it has performed in the car—it far exceeded what I thought it would be able to do! As I've said, you can buy a rear suspension from DSE that is similar to this and will work great.

Mini-Tub and Sheetmetal Tuneup

A Pro Touring car requires mini-tubs and solid sheetmetal!

If there is a 'must have' when building a Pro Touring car that will be driven hard (whether it will be on the street or the track), it is that the body needs to be solid and able to support the needs of the vehicle. As such, every Pro Touring build I do starts with having the body stripped of all paint and filler, the rusty or damaged sheetmetal replaced and any modifications (like adding mini-tubs in the back to fit monster rubber) made to the body/structure.

Rust Repair

While I do my best to buy 'straight' cars that have a minimal amount of rust and damage to them, there are still times when a vehicle I have purchased requires replacement sheetmetal panels. Years ago, I used to do my own sheetmetal repair, but I have found that sheetmetal work is best done by someone that does it every day of the week (it is more an art than a science)—and that is not me. I can do it, but it takes much longer than it should, can be really frustrating and when I'm done, doesn't seem as straight as it should be. So I outsource it.

But finding a competent, affordable shop to do sheetmetal work has proven tough. So I have a method to get with the 'right' shop and at the 'right' price to end up with a high quality product.

Shop Shopping Around

The first step after the car has been stripped of paint and body filler is to assess any structural- and appearance-rust issues. This process consists of me putting the stripped body shell of the car on a trailer and taking it to a shop, or shops, to talk with professionals about what they recommend the car have done to it and have them work up a quote for timing and cost.

Having the car stripped before doing this process makes the quote much more 'real' than if you took the car around with all the paint still on it.

In fact, I think showing up to a sheetmetal repair shop or painters shop with a car that is still painted is a waste of time—how are they going to be able to give you a quote on the work required when they don't know what will need to be done to the vehicle to get it solid, straight and shiny? If you really are committed to having the car done 'right', then it needs to be stripped before getting quotes for it.

Body Stripping Options

There are a few different ways to get to bare metal on a car body and I have tried pretty much all of them. There is sand blasting, walnut shell blasting, chemical dipping, and the old standard, sanding/grinding. Of these, I prefer the media blasting because it has shown itself to get the majority of the paint/body filler off and causes minimal immediate and long term damage. I have had some large, flat panels on cars get warped from the sandblasting process (apparently, if the person performing the sandblasting stays on an area too long with the sand, it imparts too much heat into the panel and it can warp the panel), so I select my blasting company carefully.

The chemical dipping process does remove everything on the vehicle other than the metal, but I have had long term corrosion problems with some of my cars that have been dipped. While I can't prove it was the dipping process, this issue makes me wonder if the chemical dipping process didn't contribute to it...for this reason I have been staying away from the dipping option.

The sanding/grinding process is good at removing material on surfaces you can get to with the equipment, but it is extremely time and effort consuming, so this is not for me either.

What process you choose to strip the body on your next project is a choice you will need to make yourself: but I will tell you my last three cars have been stripped using the media blasting and I have been very happy with the results.

A Local Shop

For the last few cars I have built, any sheetmetal rust that needed be to cut out and replaced was handled by a shop near where I live: Reiter Metalcraft in Mount Clemens, Michigan. While Reiter does build complete cars from the ground up, they started out doing sheetmetal work for local car builders…and sheetmetal repair is what they are known for because they are absolutely great at it. Whether it is replacing a rusty rear quarter panel or modifying the rear wheelwells to add 3 inches of inside width to them (they did a stock appearing 'mini-tub' job back when nobody knew what that was), the Reiter team just dives right in and does an incredible job.

In my experience, there are only a handful of sheetmetal repair shops like Reiter around the country. To find a similar shop in your area, you will need to talk to local enthusiasts, visit shops in person and check the local automotive enthusiast magazines to find them because what I have found is the good shops don't need to advertise: they have too much work already.

I always go and visit a shop a few times before having them do any work for me to see if the shop is clean and neat, if the cars I saw the first time are still there and if there has been progress on them. This is also good to get to the know the people doing the work.

Is this time and effort consuming? Sure, but I am asking them to do one of the most important steps in building my car, so I need to do the work to know they are the 'right' shop for the job before committing to them.

Fixing sheetmetal is a time consuming, labor intensive job, so it is expensive. The stakes are very high when you start cutting up the sheetmetal of rare, or just old, cars, so you are paying for a 'confidence' level with this work. I have heard way too many horror stories about shoddy shops ruining high-end cars because they started a job they had no business getting in to.

My recommendation is that if you can't pay for quality work right now, save your money until you can: why do everything else to your car if you can't get the foundation right on it? And in my opinion, you can't do sheetmetal work after everything else is completed, it never works out.

As for the Mule Camaro, it turned out the body was not as straight as I thought it was: it needed a new right rear quarterpanel, a patch panel in the left rear quarterpanel, both rear fenderwells (inner and outer), two front fenders and a few other small patch panels. The floor was solid, which made me feel good, but I was a little stung by how much metal was rusty for a supposed California car. Oh well, that's what you get sometimes with an older car!

Once the items to fix were identified, Reiter dove in: the right rear fender was cut off and before replacing it, we did the mini-tub rework so I could shoot pictures of it for you to see.

An importat note is that, Detroit Speed & Engineering (www.detroitspeed.com) now offers widened inner fender tubs they have stamped from one piece of sheetmetal. So on the vehicles DSE offers Deep Tubs, you just need to install them in the vehicle. DSE makes Deep Tubs for Camaros, Novas, Chevelles and probably will offer wider wheelwells for other vehicles in the future. The tubs come with great instructions on how to do the swap. If you're doing a car that DSE doesn't offer a set of tubs for your car, this section will help explain how to mini-tub the hard way!

NOTE: You can do a mini-tub job with the fenders still on a vehicle… we just removed this one on the Mule because it needed to be replaced.

Body Stripping Caution:

The idea of having the body stripped completely is in regards to getting a show-quality sheetmetal repair and paint job that will total approximately $10,000 to 15,000 or more in cost. If this is not the type of money you are willing or able to pay to have your car painted, then DON'T STRIP YOUR CAR of all it's paint. For the low buck paint job, a painter is just going to scuff the existing paint, smooth a few rough spots and spray over what you have (which is good enough for a lot of people).

Mini-Tubbing

You might be asking yourself, 'Why should I cut perfectly good wheelwells out of my car?' Well, it is a good question. I do it because the stock wheelwells don't allow me to fit the tires I want on the car: 335 mm wide Michelin Pilot Sport stickies.

This idea of 'mini-tubbing' the stock wheelwells started when I was building the 'Red Witch' Camaro project car back in '95. I wanted to run 'big' tires in the back, so I came up with a plan to extend the stock inner fenderwells 2 inches toward the center of the vehicle. The idea allowed me to fit these big tires inside the stock wheel opening while keeping the car looking like a stock Camaro. I have refined the idea over the years to what you see here on the Mule. After the mini-tub rework, the Mule rear wheelwells measure about 15 inches wide: which was just about right for 335 mm tires!

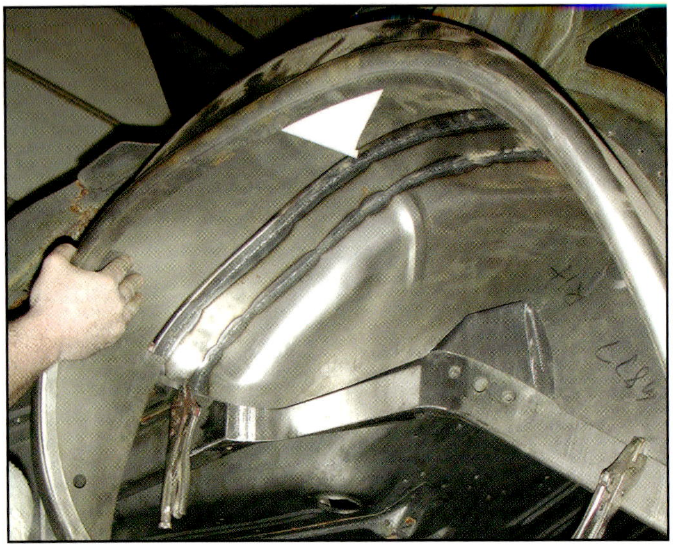

Usually, I like to have any sheetmetal repair work required on the car performed during the mini-tubbing process as they both involve cutting out sheetmetal and should be done as one of the first steps in the buildup.

This was the Mule as a stripped shell being worked on at Reiter Metalcraft. A stock inner wheelwell is being removed from the vehicle in preparation for adding 3 inches of wheelwell width via 'mini-tubbing'.

Comparing a stock rear inner wheelwell (left) with a mini-tubbed wheelwell (right) clearly shows where the mini-tubbed wheelwell space (arrow) comes from—and why the back seat bottom frame needed to be trimmed narrower to clear the mini-tubbed wheelwell!

When cutting out any sheetmetal components of a body, the pros always make sure the body is supported at multiple locations. This helps to insure the body does not sag or distort in any way. In this case, Reiter has tack welded some tubing to the back of the Mule.

As you can tell from the photo of the stock floor structure, the Mule Camaro body was extremely solid when we started cutting on it. The mini-tub process will eliminate the large step (arrow) at the front of the inner wheelwell while widening the opening.

As part of the mini-tub process, the stock shock absorber mounting pad (arrow) will be removed and the subframe 'tophat' framerail cut to make room for the big tires (near where finger pointing). The shocks will mount just inside the framerail after this process.

By sighting down the inside face of the stock inner wheelwell, the required trimming can be established to fit the mini-tub in the car. A section of the stock tophat framerail that is in the way will need to be cut off and closed back up with a filler plate that is welded in place.

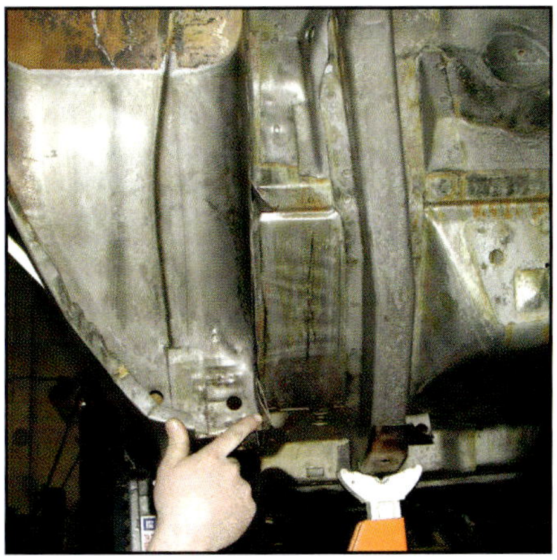

These flanges are where the stock outer and inner wheelwell mate. Adding the 3 inches to the inner wheelwell tub will move the inner flange toward the center of the vehicle (closer to the framerail).

In case you aren't familiar with the 1967-81 Camaro body structure, it is what most would call a 'unibody' construction: the frame for the body is a piece of sheetmetal stamped into a 'tophat' shape (upside down) and then spotwelded to the body floor along its length. The front frame section is bolted to the floor.

Well, here is why you mini-tub: 335mm rubber. I like to run my cars hard, so the more rubber, the better. These tires are some of the best and make these cars stop, turn and go faster than most people have the ability to drive!

To start the mini tub process, Reiter drills out what seems like hundreds of spot welds that are holding the wheelwell tub in position on the body. They use a small 1/8 inch drill first to create a 'guide hole', then drill out the remaining portion of the spotweld with a 1/4 inch drill. This is tedious work, but is the only way to get the sheetmetal out of the body!

There are going to be spotwelds everywhere, so if you have drilled out all the spotwelds you can see and the inner wheelwell still won't come out, you need to find the other spotwelds. Once all are drilled or cut, the wheelwell will still probably need some coercing to pry it out of the body (this photo shows it out of the vehicle as proof that they can be removed!).

The next step Reiter took was to trim the framerail section at the back of the wheelwell opening. This was performed with a sawzall and a long blade. This is a somewhat messy job that needs to be performed carefully to avoid cutting through the framerail.

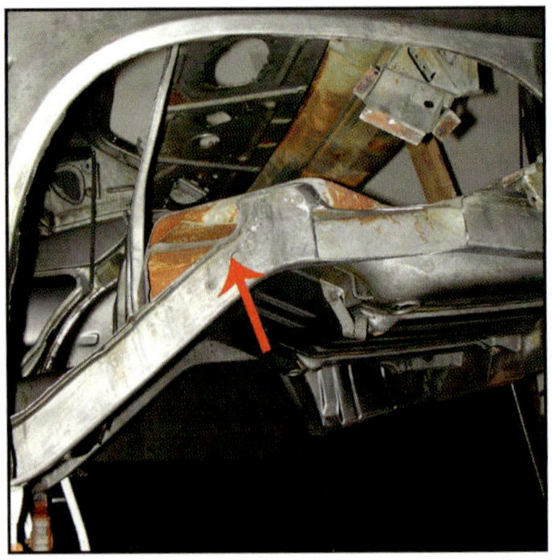

Also trimmed off at this time with the sawzall was the sheetmetal (arrow) on the outer edge of the framerail (the stock shock mount area). This area was then ground smooth to provide a clean mounting surface for the widened inner wheelwell to be welded up to the body.

There are multiple brackets that are spotwelded to the top of the inner wheelwell (arrows). These spotwelds will need to be drilled out and the brackets modified to mate to the widened inner wheelwell. A skilled sheetmetal shop should not have a problem accomplishing this.

This, my friends, is sheetmetal mastery. How Reiter does this is an entire book and you still would not be able to do it. The simple explanation is they cut the wheelwell just inside the 90° flange, build a 3 inch wide 0.047 inch thick mild steel sheetmetal insert that matches the cut line, then butt weld the insert in place without warping the wheelwell.

Reiter then created a few 0.060 inch thick closeout panels to fill the frame section that was cut out and…

These panels were then fitted and welded in place with a MIG welder. The welds were then ground smooth using a small rotary grinder to end up with a framerail that looks 'factory' built.

While the wheelwells were out of the body, the rusted/damaged left rear quarter panel lower was repaired. This required cutting out the inner panel and replacing it with a CARS replacement panel, then doing the same with the outer patch panel. The panels were welded in place with a MIG welder through holes drilled in the panel to replicate the spot welds on the original panel(s).

The Reiter-widened inner tub was then test-fit in the now 3 inches deeper opening in the floor/frame sheetmetal.

This is the payoff for excellent measurements, quality cuts/grinding and judicious welding. Notice how the trunk floor flange (arrow) mates tightly to the widened wheelwell.

The inner tub was then positioned similar to the original in the vehicle and clamped tightly against the flanges with the drilled out spot welds (arrow). A MIG welder was used to apply weld in the spot weld holes, being careful to minimize the heat in the sheetmetal (to avoid warpage).

The MIG welding was continued around the perimeter of the inner tub, including the rear frame section area (arrow). On the frame section, the flange was directly welded to the frame in stitches (in 1 inch or less lengths).

On the '69 Camaro, the inner wheelwell has a trunk lid hinge bracket that connects to it on top (like many vehicles with trunks). The deeper wheelwell requires a change to the bracket shape—here it is being test fit to understand how to change the flanges.

This shot shows how the bracket flanges (arrows) have been modified to allow the lower portion of the bracket to mate up to the widened inner wheelwell.

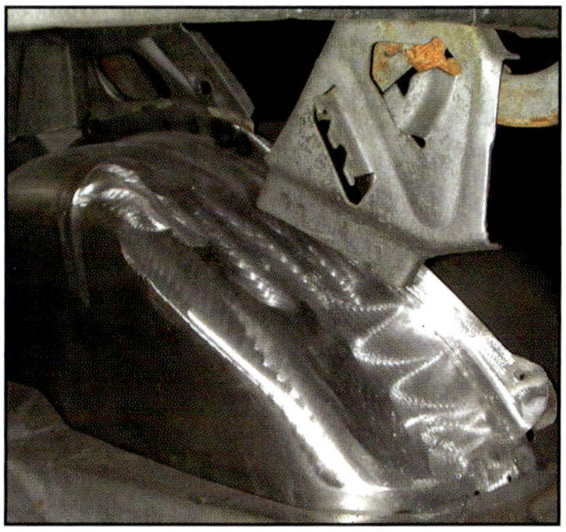

Reiter then used 60, 80 and 120-grit grinding discs in succession to grind the welds flat on the surface of the inner wheelwell before MIG welding the brackets back in position. This job is so much easier today with DSE Deep Tubs!

Deep Tubs

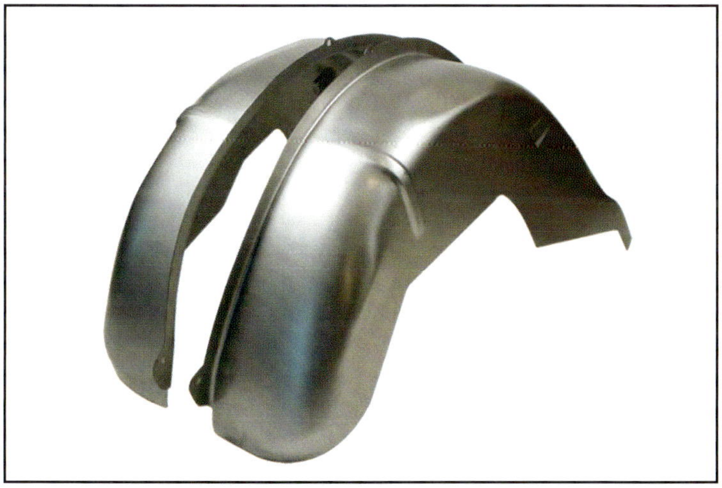

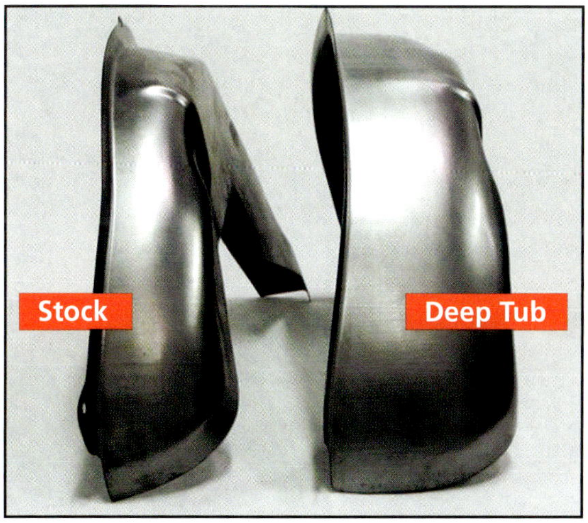

When originally doing the Mule, I commented about how mini-tubbing would be much easier if someone sold a widened inner tub: well, now Detroit Speed does (www.detroitspeed.com). DSE has these new deeper wheelwells stamped from 18 guage steel sheetmetal and offers an installation video to use for installing them. I suggest using their tubs if you can as it makes this very difficult job possible by the enthusiast (though I still would recommend having a pro install them).

Sheetmetal Tune-up

There was more sheetmetal replaced on the Mule then I like to see, but Reiter's did a great job getting it solidly installed and making it look great. Here is a summary of the work they performed with some tips from the pros. Remember that the 60's were a very tumultuous time in the US, so every Camaro body I have experienced has required custom fitment to get panels to fit and they also often have large gaps between many of the structure panels. What I am saying is don't panic if your car has ¼ inch sized gaps between the sheetmetal flanges at various points: these were often just filled (or 'skived' over, in factory terminology) with glops of seam sealer. The main goal is to get solid sheetmetal throughout the vehicle to provide structure and appearance as good or better than the original production process attained.

A reproduction right rear quarterpanel from CARS was installed on the car after being trimmed and tuned to mate to the original body flanges. Every Camaro body is slightly different, so it helps to have an experienced sheetmetal repair shop doing the installation.

Reiter likes to replace the outer rear wheelwells when mini-tubbing a car as they often are questionable from a rust perspective. So the Mule got new wheelwell outers during the mini-tubbing.

It is tough to get the C-pillar mating point of the replacement fender-to-roof to fit tightly and look good. This one is actually a good fit—it will be tuned up with leaded body solder filler to finish the install. The holes in the roof flange (under finger) were from drilling out the spot welds on the original rear quarter panel.

The mating point of the trunk filler panel to the rear quarterpanel requires a solid weld and grinding to make it all look 'right'.

The front mating point for the right side rear quarterpanel is critical for door fitment and the stiffness of the overall body. This photo should be in a textbook on how to do it right.

This is the mini-tub wheelwell job and new sheetmetal rear area of the Mule. This area will never been seen again, but the now almost 15 inch deep wheelwell area helps the Mule look like no stock '69 Camaro ever looked!

While I am all about doing whatever to save a few pounds. Reiter is adamant about using leaded 'body solder' on the rear quarterpanel to roof mating point as it has excellent adhesion, minimizes cracks and provides some strength to the body. Reiter has years and years of experience doing this, so they make it look easy, but I know it isn't.

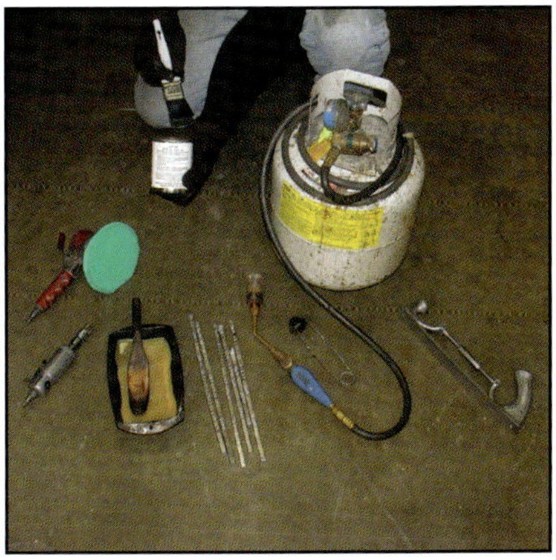

The equipment required to do body soldering includes (from left to right): grinding disc, and wire brush, wood paddle and tinning 'butter', soldering tallow, body solder sticks, leading torch (hooked up to propane tank) and a leading file.

After grinding the area clean of paint and wiping down the area with tallow to clean it, Reiter heats a body solder stick over the area they are preparing to fill until it is melting.

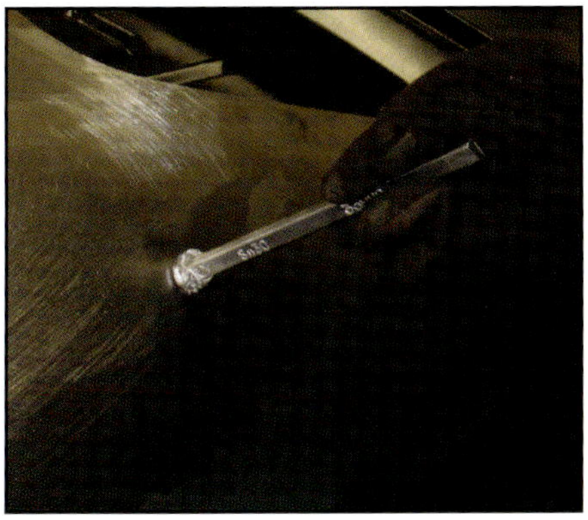

Reiter then applies the softened solder to the surface while keeping direct heat on the stick to build the area up with solder. They then use a wood paddle to push the soft solder around to evenly fill the area: being careful not to stick the wood to the solder or get the wood so hot is starts to burn (not easy to do!).

Once the solder has cooled, a solder file is used to smooth the area. Reiter is adamant about never using a grinder to smooth the solder: only use the file. This area is left for final smoothing with body filler before paint.

This is what the left rear quarter panel area looked like when Reiter was done with it: you can barely tell they stitch welded a patch panel in the bottom rear edge of it. That is talent!

Firewall Blockoff

The final mod we made to the Mule was a firewall blockoff plate to cover the stock Heater, Ventilation and Air Conditioning (HVAC) openings in the firewall. I had Reiter's do this as I use a Vintage Air HVAC system in my cars and that system doesn't need this opening. DSE now offers this blockoff plate.

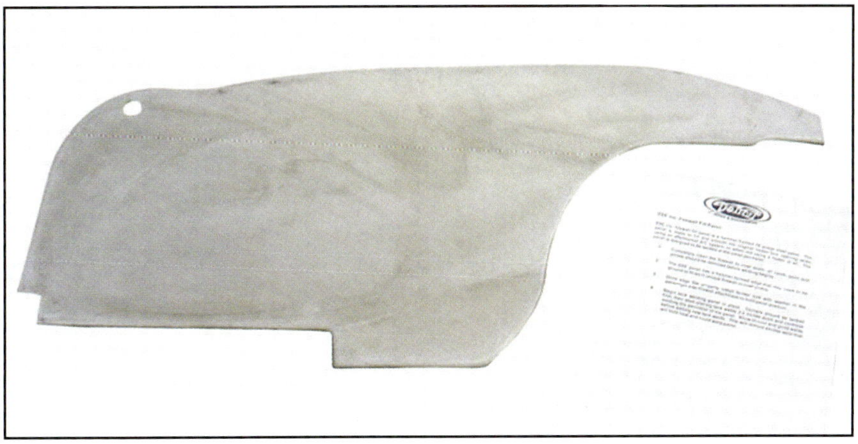

Fill Plate

Another product that Detroit Speed now offers is this Firewall Fill Plate to closeout the stock HVAC opening in the firewall. The DSE Fill Plate retains the stock cowl to firewall seam, is made of 18-guage steel, has a rolled edge and needs to be welded into position. It comes complete with installation instructions and is stamped with pride in the USA.

Lightweighting!

How to keep the weight at the bare minimum without any sacrifices

I like to have my Camaros as light as possible. When I say that to most people, they think I am all about using exotic, expensive materials at every turn to reduce the vehicle weight. While I have used things like carbon fiber hoods, etc, to keep weight down, I think the real opportunity lies in doing the basics: just keeping your eyes out for removing excess weight.

To me, the simplest form of 'excess weight', is any threaded fastener that shows more than 2 threads when fully installed. The ARP wheel stud shortening sequence shown here is a good example of that: I had to go with a longer stud to accommodate the Fikse Wheels and Baer Brakes. No problem, right? I order GM-style ARP wheel studs (pn 100-770B) and install them—but, they showed about ½ inch of thread with the lugnuts fully installed.

To save the weight, I shortened all 20 of them and avoided carrying that weight around—and by the way, they are the worst kind of weight, 'unsprung weight' on the suspension—so I really wanted to get them to be as light as possible.

I do this on EVERY fastener on my cars. It takes a lot of time and requires a lot of thought to 'see' the opportunities, but I know it helps the cars handle better when driven in anger and makes them more enjoyable to ride in down the road.

As a caution, I work very hard not to sacrifice safety or functionality with my weight savings. To me, there are more than enough opportunities for reducing weight without having to introduce other potential problems into the mix: there are plenty of those with a one-off built car already!
So, my suggestion is to always be on the lookout for ways to reduce the weight of your cars and to take action to 'cut the weight'!

Recoating - I have all of my 'shortened' fasteners stripped & recoated to minimize the chances of them rusting in the future - DETAILS!

Common, Simple Weight Savings Ideas

1. *Cut Fasteners to show only 2 threads beyond nuts*
2. *Drill 'lightening holes' in brackets where possible (without compromising strength, durability, etc)*
3. *Minimize wiring lengths between components*
4. *Use lightweight materials, if possible (aluminum in place of steel, plastic in place of aluminum, etc)*
5. *Delete unneeded accessories (power anything: windows risers, antennas, etc)*
6. *Substitute more advanced design components for stock pieces (engines, transmissions, etc)*
7. *Be creative—lightweighting is one of those aspects of car building that has a lot of opportunity in it!*
8. *Eliminate components/systems if they are not required for comfort or performance*

I lighten every part on my cars: and the number one weight savings method is to shorten all the fasteners so they show only 2 threads past the nut: here is that process as applied to the wheel studs on the Mule.

To get the correct length fastener length for the application, I have found the best way is to install a threaded fastener and mock-up the exact package you plan on running. I have threaded a nut on the stock stud and am knocking the stock stud out the back of the wheel flange in preparation for installing the ARP stud.

I then install the ARP stud and lightly knock it into the hole (you don't need to fully seat it now, just get it started in the hole). Before threading nuts on the studs, I like to wipe a healthy amount of anti-seize compound on the threads to minimize the chance of the lugnut seizing on the stud (a very aggrevating moment!).

I then place a flat washer on the stud and hand start a lugnut on the stud threads.

Note: I generally DO NOT use an impact wrench on my studs/lugnuts (zipping the nuts on messes up the stud threads very quickly)

I do use one in this case to 'pull' the stud into the wheel mounting flange by tightening the lugnut on the stud. Stop when the stud is fully seated on the backside of the flange. I then install the wheel on the stud and tighten the nut. Measure the distance the stud extends beyond the lugnut—that, minus the length of two threads, is how much I trim off the stud.

I then knock the stud out by hitting the nut (threaded off the lug two threads). I then replicate the thickness of the axle flange, disc brake rotor hat and wheel thickness with nuts and washers and trim the stud to length (shown). I have found the easiest and safest way to cut the studs is to clamp the stud in a vice and use a rotary cut-off saw. There are more elegant ways, but just about anyone can do it this way...

I then use a grinder to 'tune up' the end of the stud. This shot is to maximize the sparks effect: usually, the stud is held down against the flat surface with a few washers shimming it to keep it flat.

I then lightly grind the very edge of the threaded section of the stud to taper the end of it. This makes it easier to start the lugnut on the threaded section and looks better than a straight-cut end.

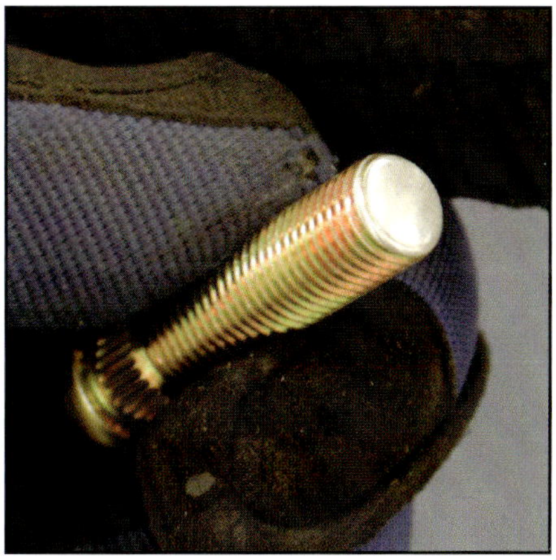

This is what each stud will look like after being trimmed—seems like an easy job until you remember there are 20 studs and way more fasteners throughout the car! Removing weight takes time, effort and creativity, but it is worth every little step.

The stock ARP stud is on the right, the shortened stud is on the left. Both would work fine on the vehicle, but one is 25% lighter—which one do you want? As a note, I send all the fasteners out for replating just before final assembling the car, as without it you get corrosion on the exposed threads. Makes a difference with the final build quality...

Subframe Connectors

Tying the front and rear suspensions together

The tri year Camaro (67/68/69) body structure is a quality design that has been used in performance situations for years with good success: but that's not saying its strength can't be improved upon. Because of that, I always installed some form of subframe connector on my Camaros to tie the front subframe rails to the rear subframe and the body structure. This improves the rigidity and strength of the overall vehicle with a minimal weight hit. This section will show the installation of what I consider to be one of the best subframe connectors on the market: the Detroit Speed and Engineering (DSE) subframe kit.

Hanging Down

I have used many aftermarket or custom-built subframe connectors for my project cars over the years and always had the same problem with them: They hang down too far. These unfortunate designs have a tendency to hit the ground at the least opportune time, limit the ability to run the exhaust and hoses underneath the vehicle and just look like crap when viewing the vehicle from the side. The DSE subframe connectors address this issue with an inventive approach, while still providing the improvement in strength and rigidity that is the goal of all subframe connectors.

The Install

These subframe connectors require cutting a small section out of the floor in the rear seat footwell to package the connectors into the floor while tying everything together cleanly. While cutting into vintage Camaro sheetmetal might seem scary, that really isn't the case with the DSE kit as it comes with excellent templates that, when followed, provide easy to follow cut lines.

The following photo sequence will give a step-by-step rundown of what it looks like to install the DSE Subframe connectors (available at www.detroitspeed.com) in a 1969 Camaro.

Installing subframe connectors is a required step if you are building a serious Pro Touring car using a subframe car (like the tri-year Camaro).

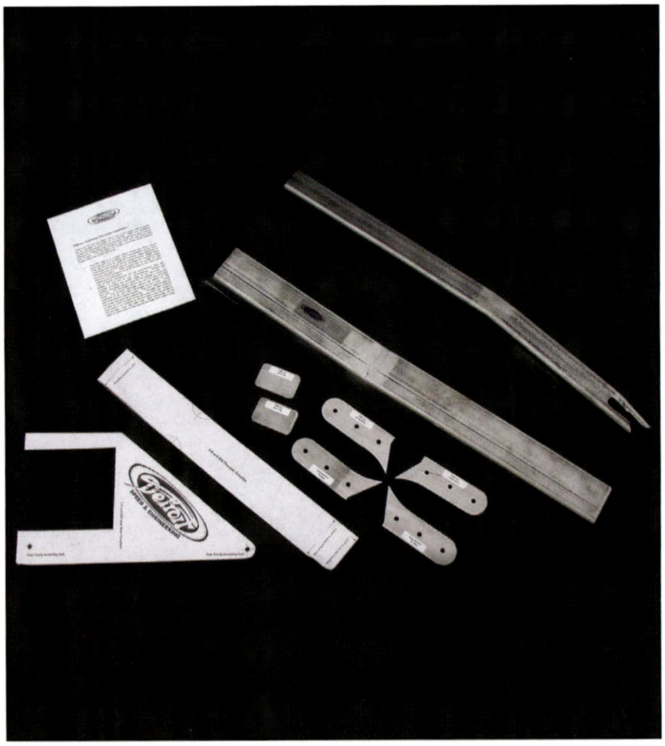

This Detroit Speed & Engineering (DSE) subframe kit (pn 010101) was one of the first products they offered. This kit integrates cleanly into the vehicle and helps to increase strength and rigidity.

Checking 'Square

The first step when you are going to tie the front and rear subframes together is making sure the vehicle is 'square'. I suggest you start this process by setting the vehicle level on four jackstands and then measuring that the wheels are equally spaced from each other on the driver and passenger sides.

Then, measure between two agreed upon points on the front and rear suspensions on each side of the vehicle to make sure they are 'square' (straight and diagonally front to back): in this case from the front crossmember to the forward leaf spring mount on the body.

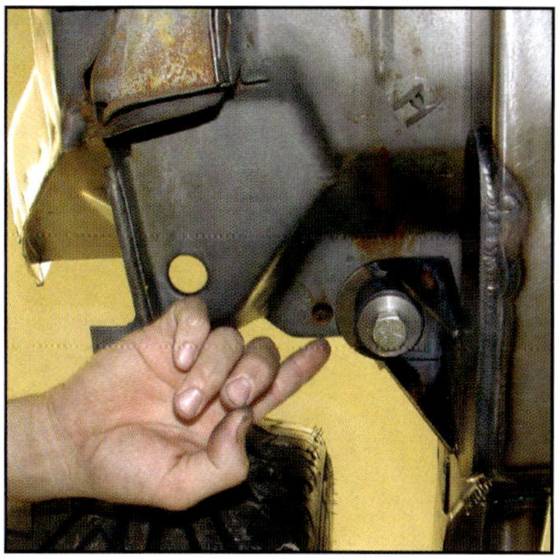

If any discrepancies in the dimensions are found, they can be adjusted by loosening the mounting bolts on the front subframe and attempting to move the subframe around to get everything 'square'.

A final measurement rechecking the position of everything should be made before the installation of the DSE subframe connectors is begun to make sure the vehicle is square.

Cutting Hole For Connector

This is the stock floorpan of the Mule in the rear seat footwell. The DSE subframe connectors will be welded into this floor section.

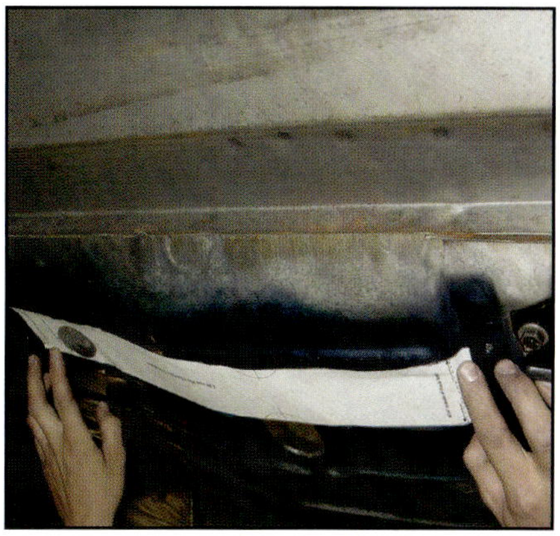

DSE has pioneered using simple paper templates to help confidently make cuts into the body to install their products. This template shows where the floor will need to be cut. Note the blue aerosol layout fluid that has been sprayed on the sheetmetal so a scribe line can be made and clearly seen during the cutting process.

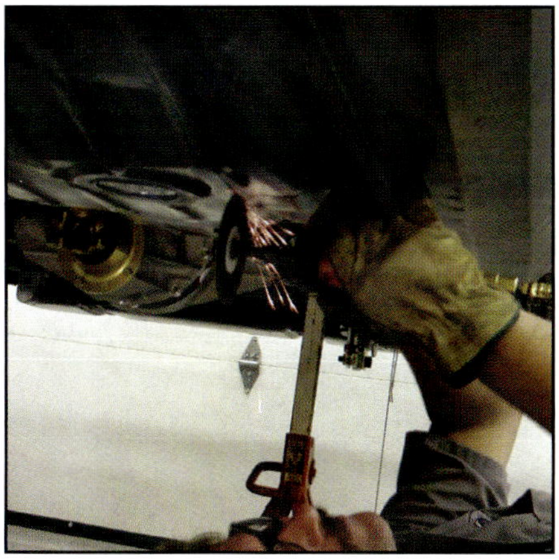

Once the scribe line has been made to the sheetmetal, a rotary cutter is used (with the proper safety equipment: safety glasses, gloves, etc) to make a shallow cut line through ONLY the lower floor panel—note how little of the rotary blade is in the sheetmetal!

There are a few spot welds that will need to be drilled out to remove this lower panel: I use the spot weld drill from Eastwood (pn 11279), but you can use 1/8 and ¼ inch drills to free the panel.

This lower panel might not come out from your initial shallow cut. If a little cleanup is needed as you peel it out, just lightly trim the edges with the cutoff wheel like this.

This is the section you should have cut out. Notice how the front section was removed from the top panel without cutting through—this is done by only cutting 1/16 inch deep with the cutoff wheel.

Here is the same cut area from an opposite view. The top panel to the left will be cut out now following another DSE template.

Once again, blue aerosol layout fluid is sprayed onto the sheetmetal in prep for marking the metal with scribe lines transferred from the DSE template.

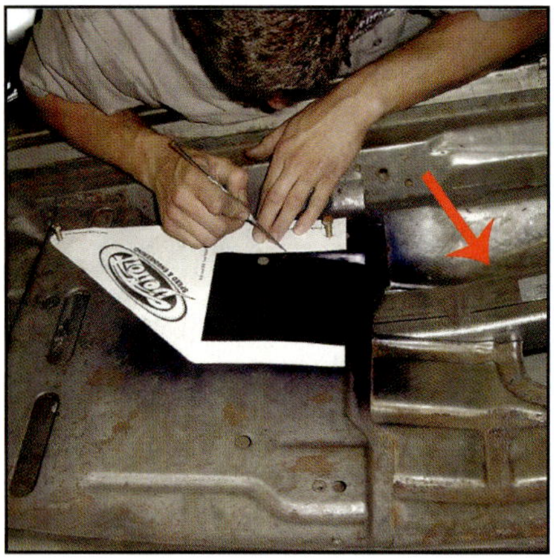

Here are the scribe lines being made using the DSE template as a guide. Note how DSE uses the seat mount bolts as a locator.

The paper template requires being held down firmly while you scribe, but it is pretty easy to get straight lines. Note how the DSE subframe connector has been 'mocked up' in the cut out area of the floor (arrow)—a mockup always helps to understand how parts go together).

The rotary cutoff wheel is again used to carefully make a shallow cut through the top sheetmetal panel, following the templated cut lines from DSE.

A few spotwelds located on the front flange need to be drilled out to remove his section from the floor. The way I find the spotwelds is to look for ¼ inch diameter depressed circles in the sheetmetal: they were formed when the factory 'spot-welded' multiple body panels together to manufacture the car.

Here is the forward panel removed from the vehicle. The remaining sheetmetal on the floor (arrow) was slightly tweaked up to allow the DSE Subframe Connector in position—this peeled up lip will be trimmed off where it touches the Subframe Connector: a weld will be made at this edge.

Openings on both sides of the car should be created, so the other cutting/trimming is done now. I have found the best way to work with a cutoff wheel is to create a 'guide groove' in the sheetmetal by lightly grinding where I want to cut, then going back over that groove a little deeper to make the cut.

Again, you will need to drill out the few spot welds at the front of the panel you just cut out to remove it and set it aside (don't toss it as this will be welded back in the car!).

Here is what the opening should look like after cutting out those two panels.

To fully mock up one of the DSE Subframe Connectors, start by sliding it rearward from the front edge of the opening towards the back of the car. This engages the 'V' opening of the connector into the floorpan.

If you did the cutting 'right', the opening should be just a little too tight at a few spots. Mark these locations, lightly grind on them, then recheck, mark, grind, recheck—do this over and over again until the connector fits into the opening—this provides the tightest fit possible for optimal strength, fit and finish.

This is one of the many times trying to fit the DSE Subframe Connector fully into the opening in the floorboard of the car. The key here is to slowly sneak up on this fit. If you need to take a break to keep your patience, do it!

Go underneath and check how tightly the DSE Subframe Connector 'nests' against the rear framerail. The connector's rear edge needs to touch the front edge of the framerail. If it isn't, see the next step (off fits are not uncommon: these cars were built in the '60s!)

One way to get the subframe connectors to fit tighter against the rear framerail is to trim two slots into the floor (arrows) until the DSE connectors fit tightly.

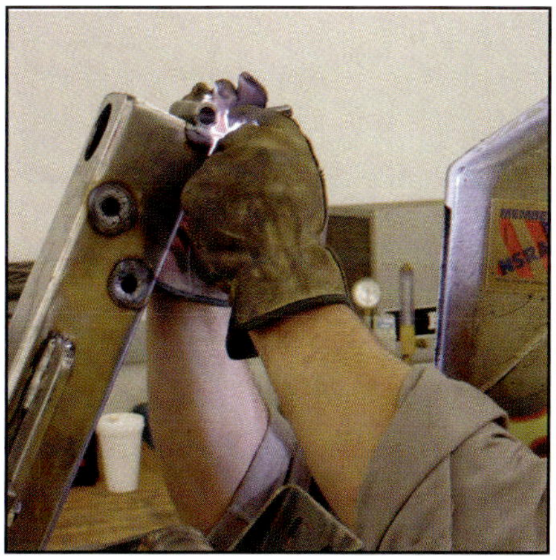

For the Mule, a Wayne Due front frame section was used and I wanted the DSE subframe connectors tied to the front frame: so three ½ diameter metal tubes were welded to the end of the DSE subframe rails so the two frames could be bolted together with some custom side plates once everything was installed.

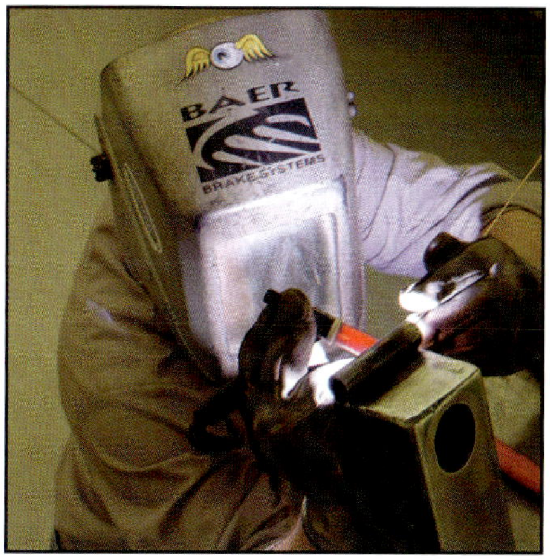

Notice how the TIG welding tip is glowing red in this photo—that is what you call putting some heat to the weld! That's also what you want on a suspension or frame component: a penetrating weld that will hold the two parent metal components together through thick and then!

Here is what the tubes looked like after the welds were ground flat on the sides of the DSE subframe rails.

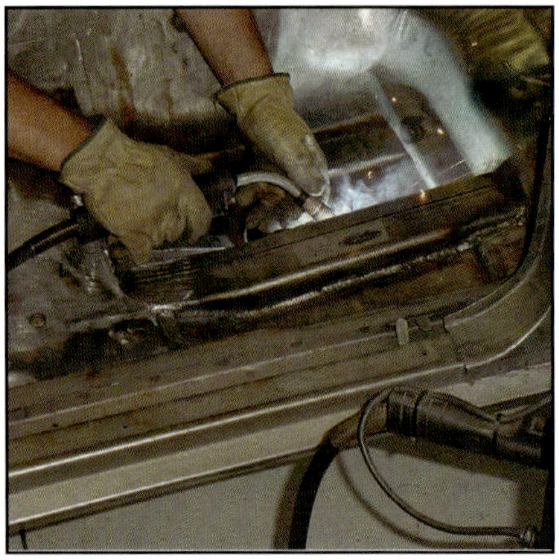

With that, the DSE Subframe connectors were final installed, tack welded in place with a MIG welder, remeasured to make sure they were located correctly, then MIG welded into position in short 'stitch weld' spurts to minimize the chance of warpage from excessive heat.

After the DSE Subframe connectors are final welded in place, there is still some work to be done to complete the job.

First, the floor section that was cut out of the interior is laid back in position to trim around the DSE subframe connector. A magnet is used to hold the section in place and a Sharpie pen guided by a ruler is used to mark the width of the area to be cut out.

The height of the trim required is then marked on the sheetmetal piece using the ruler and Sharpie pen.

The pen marks are then connected into a solid trim line that is cut along with a rotary cutter and then a small grinder to trim right up to the line.

The best way to trim a part back is to trim a little, then 'mock up' the part in position to see how much (if any) more needs to be trimmed to exactly fit the part. This usually takes a few shots before everything fit properly, but is how I recommend doing this step.

Here is the floor panel final trimmed in place. Notice the two small magnets used to hold the panel in position.

When initially MIG tack welding the closeout panel back into the car, I suggest you start with the front edge as it is laying on the floorpanel metal. Then, to weld the back of the panel in place, wedge a small screwdriver between the closeout and floorpan and lightly pry on it to position the closeout while you apply additional tack welds.

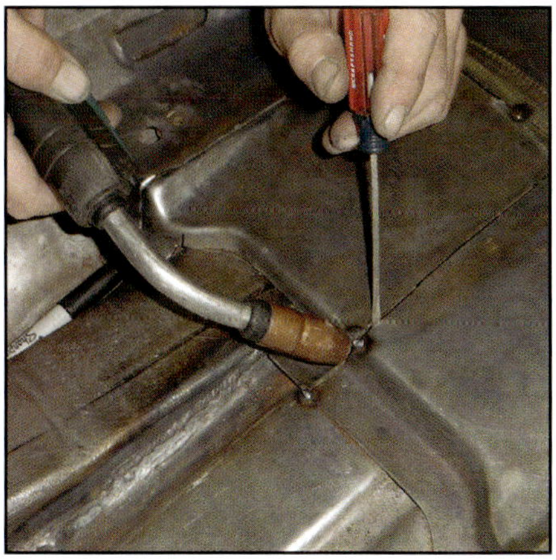

Keep moving around the perimeter positioning the closeout panel with the small screwdriver and applying MIG tack welds.

This metal is very thin, so I suggest you use tack welds (very short bursts of weld energy) to minimize the heat being imparted to the two panels and get everything welded back together—without blowing huge holes in the parent metal.

Continue applying tack welds until the panel is fully welded back in the car, then grind the welds flat.

This is what the DSE Subframe Connectors should look like when final installed.

Remember those three tubes I had installed in the Wayne Due subframe? Well, here are the side plates before being cut (notice the use of clear acetate to create the template, which is traced onto the metal plate for a cut line) out of 1/8 inch thick mild steel plate.

Here are the custom side plates welded to the DSE subframe connectors and bolted to the front subframe. Additional welding is being performed on the floor to weld in a 'tie-in' box for the rollcage (covered later in the book).

This shot gives you a good idea what the DSE Subframe Connectors do: they truly connect the framerails of the front subframe directly to the rear framerails (in this case tophat-shaped sheetmetal sections spotwelded to the floorboards of the Camaro body).

This is why the DSE Subframe Connectors are worth the effort--the low angle visuals of the car are clean AND the car has additional structure in it. The DSE Subframe Connectors are more involved to install than the others on the market but the looks, packaging and performance make them more than worth the effort.

Front-End Choices

How to set your Pro Touring front subframe and suspension up

When the Mule was being built, there were not a lot of fabricated front subframes that featured good suspension systems—most used the 'Mustang II' front suspensions that I would say are not suitable for true Pro Touring applications. But, there was one company doing a C4 Corvette front suspension-equipped subframe and that's what I installed on the Mule. The shop was Wayne Due's Chassis Shop, out of Washington. Unfortunately, Wayne closed the doors on his shop a few years ago—but he was truly ahead of his time with this product. It bolted right on, the suspension installed as though on a Corvette and it looked good. Not sure what happened, but he can at least be proud of his product.

Since the time of the Mule buildup, there have been many other companies stepping in to the Pro Touring subframe business: for me, the last car I built had a Detroit Speed front subframe and suspension. The process I followed to get the Wayne Due front suspension fully integrated into the Mule is worthy of showing, as I think what I did applies to any subframe/front suspension you would use, so here it is.

Adjustability

As with the rear suspension and other aspects of the Mule, I wanted a front suspension that had quality components and packaging, but also could easily be tuned so I could get the Mule to handle speed, cornering, bumps and anything else I threw at it. Helping this was starting with a lightweight, effective and durable front suspension system like the C4 Corvette aluminum A-arms and forged aluminum uprights. Then, adding coilover shock absorbers and an easily swapped out splined swaybar system (like they use on Stock Cars) completed the package. Everything was on this fabricated suspension when it arrived except for the swaybar system, so I built that system from scratch (and show how I did it for the first time in this book).

Getting a fabricated front suspension dialed in for your desired ride height and usage needs is critical to ending up with a Pro Touring car that handles great, looks great and feels great. This section will discuss what I did to get the Mule front suspension ready to run.

If you are building a Pro Touring car and want to have the ultimate in adjustability: combining coilover shocks with a splined swaybar system really can't be beat. The coilover shock and spring system means you can quickly and easily make everything from minute to dramatic changes by just popping the coilover system out of the car, making the change (spring rate, shock shim pack, etc) and reinstalling it. The splined swaybar is just as easy: remove the two arms off each end of the bar, slide the bar out and slide in another type of bar (keeping the same overall diameter).

While this setup might seem like overkill, I can tell you a lot of my suspension tuning has been centered around building a comfortable, neutral-handling street car that can also handle big speeds and lateral-g forces on a racecourse. This suspension setup allowed excellent tuning and resulted in a very pleasant driving car that could rip your head off on a racetrack.

Since the Mule really pioneered the mini-tubbed, fabricated front subframe Pro Touring suspension setup, this section is more for someone building a Pro Touring car using a non-traditional vehicle (something other than a '69 Camaro, '70 Chevelle, etc), that needs to modify an existing suspension or subframe (example: stub-welding a fabricated subframe built for a Camaro onto a Cutlass frame). Most likely, the spring, shock and swaybar package are going to need to be 'tuned' to the different vehicle to be optimal—and that's where having a setup like what I put on the Mule will make all the difference.

Front Suspension Upgrade

Here is the process I went through when installing the front suspension:

1. *Installed fabricated front subframe and suspension (in this case, the no-longer-available Wayne Due system: see Detroit Speed for their subframe—www.detroitspeed.com)*
2. *Mount wheels/tires on suspension and set vehicle at ride height (the stance of a car is critical, so I figure out what it will take to get that 'right' first thing), then make changes required to accommodate desired stance*
3. *Check suspension geometry: On the Mule, I wanted to make a change for improved 'bump steer': so I relocated the rack and pinion assembly up, fabricated new steering arms and notched the oil pan for clearance*
4. *Install any upgrade or adjustability you desire (on the Mule, I installed a Speedway Engineering swaybar system as it makes it easy to 'tune' the swaybar)*
5. *Build steering system (shaft from rack to steering wheel)*
6. *Install/build brake system (plumb lines, etc)*
7. *Remove everything for paint/plating, then reassemble and install*
8. *Final install subframe/suspension at ride height.*

This is my starting point: a C4 Corvette front suspension corner bolted to the fabricated front suspension. I needed to make some changes to the steering system and to be able to tune the suspension to my needs. Part of that was installing a splined Speedway Engineering swaybar system.

Wheel/Tire Fitment

Before any suspension work could commence, the first step is to settle on a wheel/tire package that fits properly on the suspension and within the fenders of the vehicle. To do this, I use some 'mockup' wheels/tires that I keep around (like this old C4 'Vette stock wheel).

I determine the wheel width, offset and positioning that can be installed on the car by stacking various wheel spacers between the mounting flange and wheel, then trying it on the car. Again, I really try to keep things simple as I feel my job is get the facts documented so I can order parts.

From playing with the ride heights early in the build, I knew I wanted the center of the subframe to sit 4 inches off the ground at the center of the crossmember. I duplicate this height with the vehicle on jackstands by measuring the distance I want the crossmember off the ground + the jackstand height.

I do a lot of physical mockup when first putting the suspension together to make sure it will function: here the front suspension is being cycled up and down with a hydraulic floor jack to check how the suspension, steering and swaybar systems would work together.

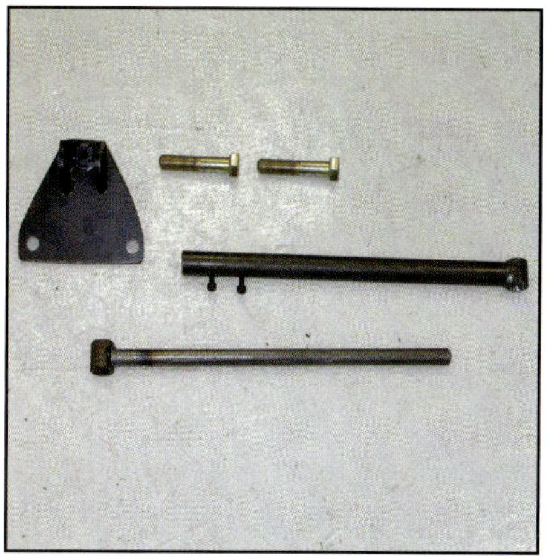

Shock Measuring Strut

Setting the 'ride height' and holding it there can be a pain, so I fabricate a 'mockup strut' that bolts in place of the shock absorber. It consists (from left to right) of a lower A-arm mounting plate, an inner/outer tube configuration with lock bolts threading into it and two through-bolts.

Here is the mockup strut mounted in the vehicle and the ride height being set with the two small Allen head bolts. This is not meant to hold the vehicle 'up'—it is just to be able to exactly determine the shock length needed for my ride height, full jounce (all the way compressed) and full rebound (fully extended).

I scribe a line on the inner rod when I have settled on a ride height setting, so when the strut is removed from the vehicle, I know exactly where I had the strut positioned. You can really see the small Allen head bolts I use to lock the rod length.

If you haven't noticed yet, I am very visual in the buildup process. To that end, here is a mockup of the actual wheel/tire/suspension/brake package on the car, at ride height. Notice how the fender ever so slightly 'winks' over the Michelin tire/Fikse wheel combo: that is the look I am going for with my ride heights.

Bump Steer Tuneup

I am very particular about having minimal 'bump steer' in the front suspension, as this really makes a vehicle feel unstable at high speeds. This front subframe needed the rack & pinion raised .875 inches to fix a small issue I found—to do this, I had to notch the front of the engine's oil pan (which was a pain).

I also work hard to maximize turning clearance, as it's very aggravating when 'performance' cars have such oversized wheels/tires that you can't even turn around on a big parking lot. This front subframe has excellent tire clearance—you'll see how I packaged the swaybar system later in this chapter to maintain the tire clearance.

Bump steer is measured with a 'bump steer' gauge that measures how much the steering hub 'turns' when the suspension is run up and down with a hydraulic jack. Here I am getting ready to do bump steer testing on the driver side of the front suspension. Part of the issue I had with bump steer was caused by my low ride height choice - I went 2 inches lower than the system was designed for.

Jumping ahead to my solution: I built bolt-on steering arms for the C4 Corvette uprights to move the steering links down approximately 2 inches. The stock steering arms will be cut off the uprights in a little bit.

I cut off the rack & pinion mounts and built new mounts from 1/8 inch thick 1018 cold rolled mild steel plate (bending them 90° in my bench vise). I then TIG welded the brackets in place to complete the fix of the bump steer.

One of the things I really liked about using the C4 Corvette front suspension uprights is that they have a fitting on them to install a wheel speed sensor. This would make it possible to run ABS (anti-lock braking system) or traction control on the Mule…hmmm.

Desired Goals

I chose to install a fabricated front end on the Mule in a bid to 'step up' the performance capabilities of the front suspension (the stock front suspension was a limitation on my previous cars) and reduce the weight on the front of the car. The Wayne Due front end did both as it utilized C4 Corvette front suspension pieces with a custom rack & pinion steering system and the entire front end weighed 125 pounds (!) less than the stock front end assembly. Needless to say, I was excited about using it on the car. There were some things I changed on the front suspension, most notably the location of the steering rack and the addition of a Speedway Engineering swaybar system, but otherwise it bolted on and looked great from the get-go.

Common Suspension Definitions

Bump Steer = *The change in direction of the front wheels as the suspension moves up and down (as when encountering a bump or during body lean into a corner). Usually caused by the suspension arms and steering links swinging different arcs. 'Bump steer' occurs while the steering wheel is held steady and usually makes the car feel 'darty' as it moves around by itself.*

Unsprung Weight = *The weight of the suspension that is not held up by the springs of the car. This usually includes the spindle, brake rotor, hub, caliper, wheel, tire and other components attached to those pieces. Half the weight of the shock absorber, A-arms, and springs are considered unsprung.*

Splined Swaybar Install

The install of a Speedway Engineering splined swaybar system would have been much easier had I not wanted to keep the original turning clearance—here you see me 'mocking up' the swaybar system and drawing lines where I plan on cutting out the frame to make room for tire clearance.

Here is what I ended up with as the system for the Mule swaybar system: A 1.125 inch outside diameter Speedway Engineering gun drilled swaybar and billet aluminum swaybar arms, Corvette swaybar links and custom A-arm swaybar link mount brackets (on A-arm). Note the bends in the swaybar arms…

This gives you an idea of how I start the install of a one-off system: I take all the pieces and somehow affix them to the vehicle (clamp, tack weld, wire tie…basically whatever it takes) and see how it goes…

Once it is all held in position, I just look everything over and start working at getting all the systems to work together.

While the swaybar system is mocked up on the subframe, I run the suspension through its travel using a hydraulic jack. This shows me whether I have at least 3/8 inch clearance between all the parts—notice I have removed the spacer I had under the swaybar frame mounts to level out the swaybar arms.

OK, I can get a little extreme sometimes in the development process: as shown here with the many different ideas I tried for swaybar links. In the end, I used the factory Corvette C5 swaybar link (on the far left) as it is the nicest: investment cast, sealed ball ends and plated surfaces everywhere.

Detroit SPEED, INC.™

Hydroformed for the Future

Detroit Speed offers what I consider the best front subframe in the business: It has excellent suspension geometry, hydroformed frame rails and stamped crossmembers, It utilizes C6 Corvette knuckles and bearing packs and comes equipped with rack & pinion steering, coilover shocks and a splined swaybar. While you can order it in bare metal or powdercoated, I suggest you get it powdercoated as it saves a TON of time and hassle!

I cut out a section of framerail, then built a fillerplate to 'box' the section closed. I then tack-welded the plate into place to end up with this photo.

With the basic system ready to go on the car, I began the process of notching the frame so I could tuck the swaybar arms close to the frame. Here, I am measuring my turning angle to make sure I can package the swaybar arm in the vehicle without losing any turning radius.

I cut the framerail to its center after determining where the swaybar arm would be positioned, then bent up a closeout plate from 1/8 inch thick mild steel to cover the opening. Here, I am TIG welding the plate into the framerail.

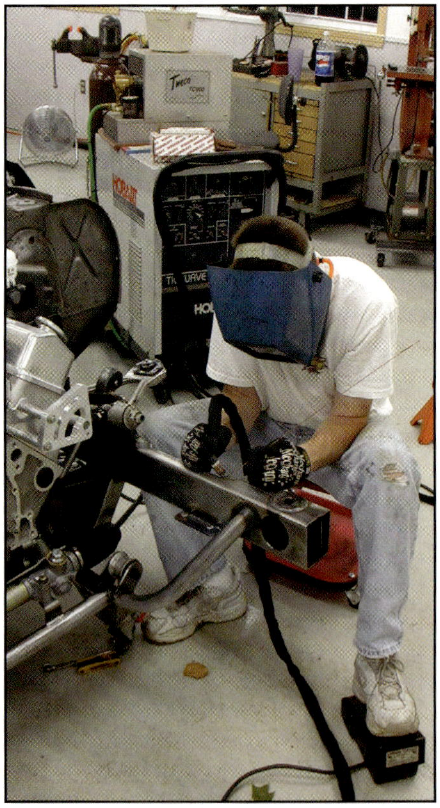

Welding like this takes some time as you don't want to put too much heat on the rail & warp it, but you need enough to get a good weld.

This is what the framerail area looked like after making the change. I also welded a small pad onto the bottom of the framerail that is exposed on the inside of the framerail for use as a mounting point for the swaybar.

The C5 link was attached to small brackets I built out of 3/16 inch thick, 1018 cold rolled mild steel plate and mounted to the lower A-arms. Then, I positioned the links vertical with relation to the car (in both front-to-back and side-to-side positions) and drilled a hole in the swaybar arm. The swaybar arms need to be bent to hold the links vertically in position.

Getting all the pieces mocked up so they work well together takes lots of time and thinking, but it is one of my most favorite times of the build. I like to keep my shop clean and neat during these moments to help me stay focused on the job at hand—you'd be surprised how much this helps!

To start the process of bending a swaybar arm, I clamped one end of the arm in my large bench vise, then clamped a long piece of bar stock to the other end. I then applied an even amount of heat to both sides of the bend area on the swaybar arm with a 'rosebud' acetylane torch and began lightly pulling on the bar stock. Once it was soft enough to bend, I pulled just a tick past my desired bend point and let the arm spring back to my final position.

The arm on the top is just out of the vise after being bent. I have polished the bend area on the bottom arm to minimize the impression left by the vise (red arrow) and the bar stock (blue)—I was concerned these would be 'stress risers' and possibly encourage a crack to form in the arms.

This shows how you change the 'rate' on a splined swaybar package: the wall thicknesses are varied. The swaybar on the right has a higher 'rate', or resistance to twisting, than the swaybar on the left—but they both have the same outside diameter (OD).

I ordered 1.125 inch OD swaybars from Speedway Engineering, which is a common stock swaybar diameter. This way, I was able to order some rubber mount swaybar brackets from Energy Suspension and use a simple lock collar (installed on the swaybar here) to position.

Another step I needed to do was countersink the hole for the swaybar link in the swaybar arm so the nut would engage all its threads on the link's stud. Also notice I have cut the arm just past the drilled hole and bevel ground the end so it looks good.

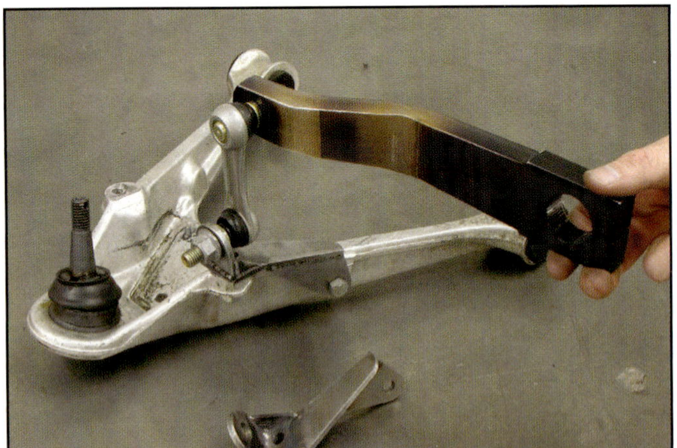

This is what the system will look like on the vehicle: looks pretty nice, huh? Its times like these when I really smile. I had an idea, got it all worked out in my head, bought the parts I could buy, modified or built everything else and put together something that works pretty good. A good day!

The Speedway Engineering swaybar system was then installed for a final check of the Mule front subframe and suspension system.

Here is why I did all the work to package the swaybar arms in tight on the framerails: this will result in a surprisingly small turning radius, provide for excellent car control (especially when I need to countersteer in an oversteer situation) and looks cool too!

Notice how close the end of the swaybar arm comes to the coilover spring/shock. There really is not a lot of room once all the components are installed on these cars!

I like to leave ½ inch of room between the tires and any hard parts as the tire sidewalls can deform during hard loading—but usually not more than ½ inch. Notice the two bolts visible on the mounting pad for the swaybar bracket.

The Mule front suspension was really ahead of its time: lightweight, independent front suspension/subframes with splined swaybars are available from multiple companies today. I am very proud of how well this works and looks.

Special thanks to Wayne Due, QA1 shocks, Fikse Wheels, Michelin tires, Speedway Engineering, Detroit Speed & Engineering, Baer brakes and ARP fasteners for helping to make this part of the Mule buildup happen!

Steering Tune-up

Build a simple steering system that works

To me, there are two systems on your Pro Touring car that MUST ALWAYS do their job: one is the steering system, the other is the brakes. If either of these two do not respond in your time of need, things will end badly. For that reason, my steering and brake systems are usually built for reliability before anything else. Yes, they still look good, but the main focus is on building systems that can handle anything thrown at them, are understated and are easy to maintain.

With that in mind, the steering system on the Mule consists of a late 70's GM steering column that I rebuilt and modified, a Borgeson U-joint and double-D shaft assembly, a rack & pinion steering assembly and custom steering arms for the C4 Corvette uprights.

Probably the most 'engineering' I usually do is get the Borgeson U-joints and double-D shaft to connect the steering column with the steering box or rack & pinion assembly. Unlike my previous cars, the Mule has a fabricated front end (from the now closed shop of Wayne Due). The advantage of this and other fabricated front subframes is they use the more modern rack & pinion system that is lightweight, provides more precise steering input and packages better into the vehicle. The packaging improvement is a big deal as the original Camaro steering box is bolted to the driver side front framerail and can be difficult to run the steering shaft straight forward to connect to it—with headers, fabricated A-arms and other components in the way.

To give you an idea how much I liked building and driving the Mule with its rack & pinion steering, every car I have built since the Mule has had a fabricated front subframe/suspension with a rack & pinion steering system! It makes a huge difference and is worth the effort.

While you probably will not have to do as much engineering as I did to make the Mule work (the many subframe/suspension companies on the market today have fully engineered steering systems from steering wheel to tire), I think it's good to understand how many of these systems go together and what to watch out for when putting them on your car.

My steering systems are pretty basic and use as much production parts as possible—as it's a system I really want to work all the time

I use the stainless steel Borgeson 'double D' shaft and U-joint combination to connect the rack & pinion steering box to the steering shaft. The 'double D' shaft makes it impossible to clock the U-joints out of phase and looks better over time (the splined shafts get goo and grit stuck in the splines over time—which looks gukky).

To connect the Borgeson double D shaft to the mid 70s GM steering column I am using, I needed to cut off the factory 'rag joint' mounting pad, then weld in a 'spud' section of the double D shaft into the steering column shaft. If you do this, make sure to disassemble the steering column or you'll melt the Teflon bushings in it.

The advantage of using the Borgeson double D shaft is that the phasing of the U-joints is 'locked' (unlike with a splined shaft where you can get the U-joints out of phase). The short piece of shaft is what I welded into the stock steering column shaft (most of this is slid up inside the steering column shaft).

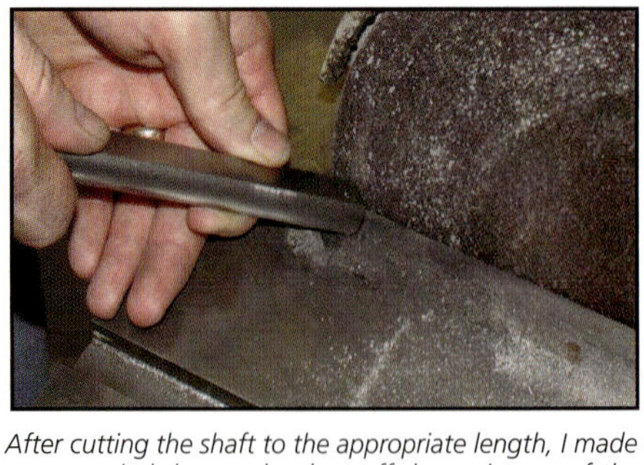

After cutting the shaft to the appropriate length, I made sure to grind the rough edges off the perimeter of the end so it would slide easily up into the steering column shaft.

Here is the steering assembly mocked up in place: the steering column (on the right) has the short 'stub' shaft slide up inside it and the U-joint is attached to the stub shaft. The long shaft (on the left) runs down to the rack & pinion assembly.

This shot shows how the steering shaft winds its way between the driver-side exhaust header primary tubes and engine mount to reach the rack & pinion input shaft. I like to maintain at least ½ inch clearance between the shaft/U-joints and components (headers, frame, engine mounts, etc) to avoid contact.

To complete the steering system, the tie rods for the rack & pinion connect to custom-made steering arms on the uprights. In case you don't remember (from p50), I built these steering arms when I relocated the steering rack up (in car position) to improve the bump steer on the front suspension.

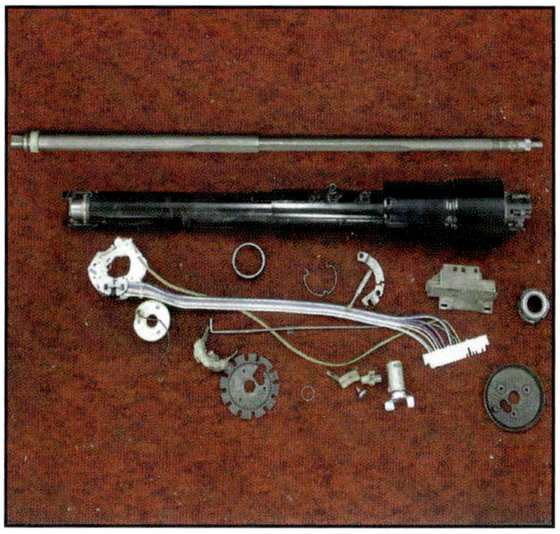

For the Mule and many of my other project cars, I used a mid 70s steering column that came out of a local junkyard. I like these columns because they are simple, look good in the car and they do the jobs required (key cylinder, steer, tilt, turn signal). I always disassemble them to do the welding on the shaft end, clean them, and paint them and disable the column lock.

After removing everything from the steering column, I pull the factory key cylinder lock mechanism, which includes the 'Rack Assembly' (it's the thing with the teeth on it) and the Locking Pin, out of the steering column so I can disable it (I do this so I can still steer the vehicle if the throttle sticks wide open and I need to shut off the engine!).

This is what the Locking Pin looks like once I've ground down the end of it. I then apply white lithium grease and slide it back into the steering column.

After fully cleaning it up and wiping it down with grease, I reinstall the Rack Assembly down into the steering column Sector Gear (that's the gear that engages with the ignition key lock cylinder).

I then check my work by putting the ignition lock cylinder into the column and turning it. As an FYI, I always make the door, trunk and ignition key cylinders operate on one key—this makes life much easier with the vehicle. Also, I put a hidden kill switch in all of my cars as another security backup.

I then install the closeout plate for the rack and slide it onto the shaft housing and start the reinstall of the column into the vehicle.

Rollcage Building

How to create a strong, unobtrusive Pro Touring rollcage

Building a rollcage for your car is not for everyone. It requires specialized equipment like tubing benders and 'fishmouth cutters', the actual process on how to use the tools needed are not well known and the consequences of doing a job incorrectly are serious. With that said, whether you are paying someone else to build you a rollcage, or building it yourself, the process should be understood.

Which brings us to this section: I build my own rollcages and use my own simplicity-based system that I am going to share with you here.

I am sure this is not the only way to do a 'cage: this is just the method I have created to end up with a rollcage that fits my needs. I also doubt it is the fastest method, but it is meant to make it easy for me to successfully complete the job of building a rollcage.

Why Build It Yourself

I started like everyone else in this business: not knowing how to do anything. So, as with most of what I have learned and done, I initially spent a lot of time asking myself what I would want in a rollcage.

The answer is I want a rollcage that:

1. Meets all racing/competition requirements for places I intend to compete at (road course, drag strip, land speed racing, etc)

2. Provides the maximum amount of protection and chassis rigidity increase while minimizing weight increase

3. Is nearly invisible both visually and regarding entry and exit of car by the passengers, and

4. Provides a solid mount for the racing seatbelts (in this case Sparco units)

These seem to all fit together nicely, but I can tell you achieving them all is tough to do—but it is possible.

How to Learn

To create my vision for a rollcage, I looked at what others did to build a rollcage and how people installed the 'cages in their cars. From all of this, I created a plan on how I would want my rollcages to look and work in my cars and then worked backwards from there to understand how to build a rollcage for my car.

I have put at least a four-point rollcage in just about every one of my Pro Touring cars: so I have learned how to package them tightly into the car, maximize the strength they add and do it myself. Here is what I know about Pro Touring rollcages. DSE offers a rollcage that is similar in design to mine.

One of the issues I noticed early on from rollcages available for purchase was they did not use the interior space effectively. Since I like to drive my cars, having a rollcage that is designed and bent to sit 2 inches from the top and sides of the interior robs considerable interior space. This was something I deemed unacceptable early on and why I decided to build my own rollcages.

One of the early realizations I had when researching building a rollcage was the need for specialized equipment: you'll need a TIG and/or MIG welder, a tubing bender & various other fabrication tools. I had a TIG and MIG welder, and ended up buying a Mittler Brothers tube bender (which I've used for many years without any issues). The Mittler Brothers unit has been easy to use, is very compact (for easy storage) and has produced great looking bends time and time again. A note on the tubing to bend: I always use Drawn Over Mandrel (or DOM) seamless mild steel or chrome-moly steel tubing—not black pipe—as many 'experts' have told me DOM has the best strength and crush properties for use to build a rollcage.

Which brings me to the last, but major issue. If you don't feel confident about your welding skills, have someone else do the welding. Welding the tubing mating points together on a rollcage with welds that have maximum penetration (for max strength) is tough to do if you are skilled and confident, so don't try to learn on this type of project.

With that, check out how I build a rollcage…

I build a rollcage by first constructing a 'template' for the entire cage to use as a guide. This is a main hoop template I built years ago for the 67/8/9 Camaro. The Mule has a 10-point rollcage (2 points each for leg: main hoop, rear frame down bars, rear spring mount stiffeners, dash downbars and front frame downbars) – note the rods going forward and back from the main hoop template.

This is why I first use the ¼ inch rod instead of working with the large tube: it is small, cheap and flexible. If I screw up with a bend in the ¼ inch rod, I just bend it around the way I need and keep going. If I cut it too short, I just pull out another length of rod and recut. Then, when the ¼ inch rod templates are done, I can store them easily for future use.

I always position the ¼ inch rod so it replicates the centerline of the 1 5/8 inch chrome moly rollcage tubing I like to use. I use 11/16 inch thick wood blocks (1 5/8 inch divided by 2, minus ¼ inch divided by two, = 11/16 inch) where the tubing is being positioned next to a hard surface.

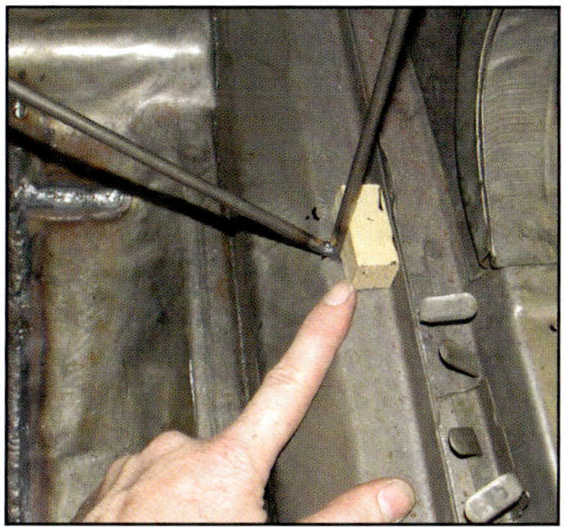

This rod & block of wood represents where I want the main hoop downbar to mate with the floor rail. Notice how the surface where the tube will mate is angled: in cases like this, I build a metal plate pad and weld the rollcage tubes to that. This way, when the weld is made, I can look at it and know it has good penetration and will provide an excellent bond.

The front downbars are probably the hardest to package of all the tubing. You want them high enough so you are not always knocking your head into them and the front portion of the tube should be close to the 'A' pillar to avoid blocking out huge portions of the road. This is all done in three dimensions and there is no room to work. The ¼ inch rod mockup is crucial to success.

There are tons of ways to cut rollcage tubing: I use a chopsaw I got about 20 years ago. It is not supremely accurate, but I 'tune' the end of each tube on my bench grinder (if it is a flat face) or with a hand grinder (if it needs to be a 'fishmouth' end to mate to another tube).

I've used a Mittler Brothers tube bender to create my last 4 rollcages and I am constantly impressed with how accurate and easy it is to use. Notice I've created a 7 inch radius plywood template of the bend mandrel (which replicates the 7 inch radius bending shoe I have for 1 5/8 tubing). The lines on the tube represent the cut line (near right hand) and where to locate the tube on the bending shoe.

The Mittler Brothers bender uses an 'air over hydraulic' system to impart force to the bending mandrel: so you will need an air compressor (I have one plumbed into my shop) to power the tubing bender. I am pressing on the actuator for the bender with my left foot while guiding the tube in the bender with my hands.

Using oversize pieces of tubing or formed sheetmetal, I have created various tubing template collars that I use to replicate a straight (shown here), various angles, fishmouth and other common end shapes. These have come in VERY handy over the years...

So here is how my system works: I create the template out of the ¼ inch metal rod (tack welded together) and then cut the rollcage tubing to length, bend it and compare it to my template.

If the tubing bend is not exactly what it should have been, I have found a few degrees can be 'unbent' using some 'cheater' bars and some weight to do so. This only works for a few degrees and only once…so if you create a 35° bend where a 20° should have been, you just built a nice piece of scrap: measure twice, cut (or bend) once!

This is a good explanation of my rollcage building process: I have final welded the main hoop in place (with an angle that 'hides' it behind the 'B' pillar mating point of the front and rear side glass) and am mocking up the various tubing that mate to the main hoop. Everything gets built outward from the main hoop…

Creating tight fitting 'fishmouth' tube ends like this is time consuming—but key to maximizing the strength of the rollcage (I recently bought a 1 5/8 inch cutting tool for my mill that makes this job much easier than grinding!). When test fitting everything in the car, I use hose clamps, safety wire and the occasional tack weld to hold the 'loose tubes' in position while I mark where to trim. Then, I remove the loose tubes, grind on the ends a little and reposition them in the car.

This rear downtube is an interesting thing. When I was running a leaf spring rear suspension, it made complete sense as one of the spring mounts was back here. But the Mule has coil springs mounted up under the rearend, so this really doesn't help the car perform better—it would probably help in a crash, though…

To 'tie' the rollcage into the load box that holds the mounts for the coilover shocks, I built this short, diagonal tube (arrow) and welded it into position. It sits directly on the top of the shock mount which is on the other side of the load box.

This is where building a rollcage on a street car gets ugly: going through the dashboard and firewall. The tubing placement is a balance between providing maximum strength and passenger comfort while minimizing the negative visual impact. I chose to run the downbars as far forward as possible and tie the front downbar through the firewall under the dash: which is a good compromise for this situation.

To get this front downbar tight against the 'A' pillar and into the dashboard without cutting the dash out, I created two separate pieces of tubing and mated them just above the dashboard by 'sleeving' them: essentially welding them together, but sliding a tube inside that has an outside diameter (OD) that matches the inside diameter (ID) of the rollcage tubing, then welding the two tubes together.

Looking at the same test fit through the windshield gives you an idea how close everything fits together—its like a jigsaw puzzle that you are cutting the pieces to as you go! I am proud of how this looks, but don't want to do this again as it is ridiculously time consuming.

OK, here is my wood block trick again (I really like to use these as they make great 'go/no go' gauges) on the 'A' pillar fitment. The interior piece is both fragile and a pain to get in and out of the car, so I placed the wood block to replicate the required opening—I also used this to hold the tube in place while welding it.

The shot isn't that great (it's tough enough to work in this space, much less get pictures of it!), but here is the front downtube that I created to run through the dashboard to the backside of the firewall. It has multiple bends, two closeout panels and was a helluva deal to fit in the car. The front suspension downbars bolt to the firewall as I wanted to be able to remove the front subframe if needed.

The 'downbars' on the front subframe were designed to bolt to the firewall--this made it possible to remove the bolt-on front subframe. Doing this was harder than I thought as I had to add threaded 'fittings' on the firewall, but once everything was massaged into place, it worked great. Most aftermarket subframes use this today.

Sometimes, two things occupy the same space and modifications need to be made. I this case, the clutch slave cylinder 'crashed' into the stiffener tube on the front subframe. So I used a hole saw to create a 'half moon' cutout, then welded a piece of tubing into the opening to close it back up and create strength.

So here is the rollcage welded into position. I am creating the removable main hoop crossbar mount that I use to hold my seatbelt shoulderbelts when I am competing with the car—and remove when I'm not so I can use the back seat.

Here is a closeup of the crossbar mount: it is a 'double shear' mount (two side plates). The bar is positioned behind the main hoop to maximize the chances it will hold me in position in the event of a crash. This level of detail is what makes a rollcage 'great'.

My Simplified Step-By-Step Rollcage Building Process

1. Study cars that have rollcages in them to determine what you want/don't want (shoot photos for reference later on)
2. Bend a mockup rollcage template using ¼ inch round, hot-rolled steel rod by hand around wooden bend radius tools. Tack weld pieces together. Mock up in car with you sitting in it. Make changes to template (this trick is from master fabricator Rick 'Speed' LeFever)
3. Weld diagonal 'stiffener' pieces into mockup to 'lock in' shape of mockup pieces
4. Build and weld in 'flat' mounts on the floor to make it easy apply penetrating welds to rollcage tubing/floor
5. Use mockup template as guide to bend up rollcage tubing
6. Test fit bent tubing in car, trim end lengths then fishmouth or otherwise tune up ends for tight mating point
7. Weld rollcage tubes into place in car

Pre-Paint Assembly

Trial fitting the major components in the car before teardown for paint

I always do a 'pre-paint' assembly of my cars before tearing them down for the components to be painted, plated, polished and/or powder coated. This takes time and effort, but I have found there are no shortcuts in this side of the build process. If you don't do this work beforehand, the chances of you having to grind off paint to weld something to the car or bash something into place are very high. Simply put, the resulting product will be less than what you dreamed all those years about—is that what you want?

Believe me, I know this is a low point in the build. You're tired, frustrated and want the thing painted so you can start the final assembly process. The right thing to do is reassemble EVERYTHING and make sure it all goes together. Make sure everything looks really good in bare metal. That way, it will look spectacular when painted/plated/powdercoated.

I strongly suggest you completely build your car 'in the raw' before sending it out for painting. Here, I am installing the entire front subframe, suspension, steering, drivetrain and steering system on the Mule to check that everything goes together smoothly. By the way, that aluminum center section Currie rearend is 13 pounds lighter than an all steel unit - which is why I use it!

Top 5 Ah-hah! Moments in Pre-Paint Assembly

1. Leave Room For Coatings

Always leave clearance for paint, plating or powdercoating when pre-assembling components. Paint is commonly 0.005 - 0.010 inches thick, plating (chrome, nickel, etc) varies between 0.003 and 0.010 inches thick, anodizing is 0.003 inches thick and powdercoating can be up to 0.030 inches thick. This might not seem very thick, but believe me, it is enough to make stuff not want to go back together!

2. Protect threaded holes

I don't like having to go back and run a thread cleaning tap down every threaded hole in the car, so I plug the holes with temperature-rated plugs (most plating and powdercoating shops use them).

3. Dress all edges

It is a good idea to inspect all the edges on the various components that will be going on your car. Don't assume your painter/plater will prep these edges—often they just coat them—jagged edges and all. I make sure all the edges are smoothed before sending them out for coating.

4. Transport Carefully

How you deliver your parts/pieces/car to the vendors applying the final coatings says a lot about how you expect them to treat your parts. I package everything in new boxes wrapped in bubble wrap and marked with my name. I expect everything to come back treated as good or better than I delivered them.

5. Inspect Everything, Pay For Good Work

Many good painters/platers/powdercoaters will make you pay them before they release the parts/car to you. I have no problem paying for good work—but I always inspect the parts/car in good light before taking possession of anything. If something is not satisfactory, you need to call it out right then, come to an agreement with the vendor on that issue and move forward.

The key when pre-paint assembling a vehicle is to account for the coating thicknesses. In this case, the bolt holes for the engine mount brackets need to be 0.010 inch oversized to allow the bolts to install after they are painted.

I also determine all of the fastener dimensions during this pre-paint assembly step: that includes diameters, lengths, thread types (coarse or fine, metric or English), Grade levels and head types. During the early build steps, I use whatever bolts I can find to get the car together—but this is the last time those fasteners will be on the car.

The Mid-Build Showing

A good way to get excited about this step is to invite your buddies/family/co-workers over for a low-key party in your garage or go to a car show with the car in its pre-paint build stage. This way, the pre-paint assembly step has even more purpose. A fun part of doing this is that people get a better idea of what you have done with the car because it is easier for them to see your changes in bare metal—believe me, there is plenty to talk about and you will have fun.

Cooling System Buildup

Maximizing Radiator Performance

I am big on keeping the complexity to a minimum on my vehicles—which might make some of you laugh, but really, its true. Take the radiator as an example of my way of thinking. I don't like to have a bunch of separate coolers hanging off the front of the vehicle (radiator, engine oil cooler, trans cooler, steering fluid cooler, etc). So, I integrate those coolers into the radiator and its side tanks (like they do on production cars), so I can still maintain the temperature of those fluids, without the messiness of the multiple coolers.

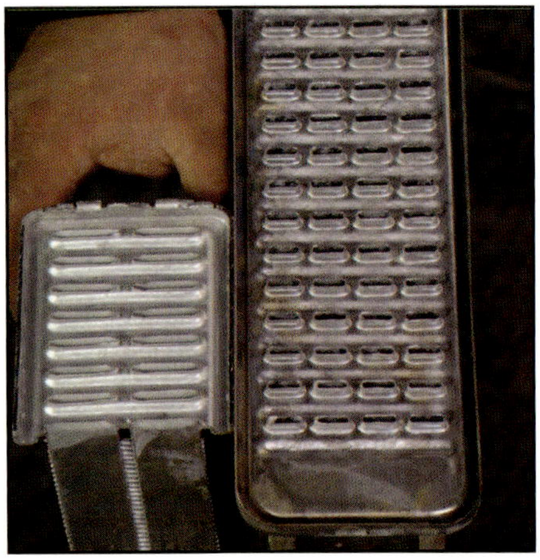

One of the choices you should consider is what style of radiator core you use: and yes, there are BIG differences in how a radiator performs depending on the type of core design you use. In this case, the dual oval tube radiator core on the left is designed for high-speed usage while the quad tube on the right is designed for more low-speed operation. I'm using the core on the left.

This radiator was was the first of a few radiators the Mule has run. I had to change them when I went to different engine packages (the 'last' engine being a 1000 hp twin-turbo small block V8). Every car I have built could sit in traffic for hours with the a/c on full-blast and not overheat—and that is the result of properly designing and sizing the radiator for the application.

I suggest you list out your goals with the radiator company (I've had good success with C&R, BeCool, Fluidyne and Griffin) you choose to build your piece, then do what they recommend to get the performance you want.

Having a custom radiator built for your application is one more way to insure the car is reliable, fun to drive and easy to maintain. Design it with help from your radiator company to get the performance you desire.

Top Balancing Points

Here are what I have found are the conflicting desires when ordering a radiator:

- *Keep it small and lightweight*
- *Have it provide more than adequate cooling for all situations*
- *Make it look cool*
- *Be low cost*
- *Have it be durable*

I use the factory mounts for products whenever possible as they seem the most robust. Here, we are creating the mounting tabs that will position the radiator on the radiator core support. Notice how big the radiator is—I want the most heat shedding capacity possible!

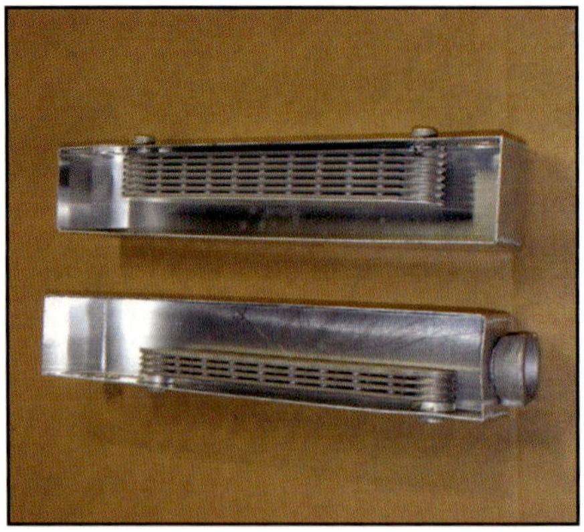

The side tanks on my radiators are loaded with auxiliary coolers: in this case, the one on the bottom is the power steering cooler and on the top is the engine oil cooler. These are available from your radiator company: but you need to take them into account with the radiator sizing as now the coolant will also need to shed the heat from these exchangers plus the heat from the engine!

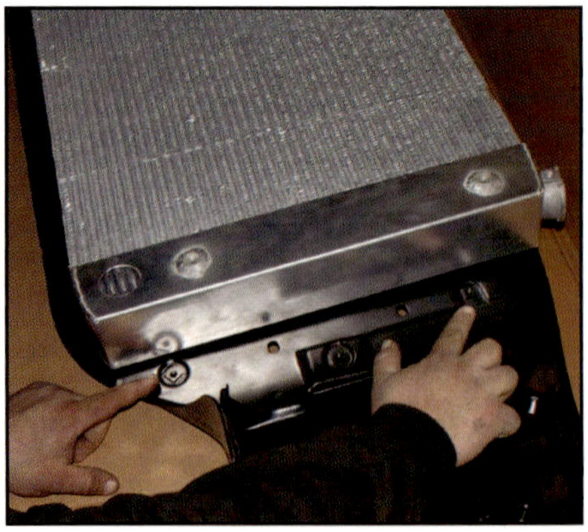

So here are the tanks mocked up with the radiator core on the core support to determine where the mounting tabs need to added. This way, when I need to install and/or remove the radiator, its as simple as draining the coolant, removing the hoses and removing some bolts.

Battery Basics

Getting the Battery Located

If you are building a Pro Touring car, then all-around driving performance is something you are striving to achieve with your car. One way to maximize performance is by getting the weight of the vehicle as balanced as possible front-to-rear and side-to-side (most musclecar-based Pro Touring cars are nose-heavy). A very simple way to assist this is by locating the 12 volt battery for the vehicle in the right rear of the vehicle (vs the original location under the hood). Most car batteries weigh 30+ lbs, so taking that mass off the front of the car and moving it to the rear makes for a sizable change in weight bias.

This might seem like a small change, but it is not—on two fronts. First, the actual mounting of the battery is relatively easy. I suggest you get a good mounting kit: the one I like to use is from Detroit Speed & Engineering (DSE). Then, find a flat area (or create a flat area) in a sealed compartment on the vehicle. For the Mule, I mounted the battery over the rear axle on the right side of the trunk. This location provides improved weight distribution (helping to offset the driver's weight).

Second, wiring the battery into the vehicle requires a substantial thickness copper wire and robust connectors to tie everything together. While the copper wire does add some weight, it is usually routed along the bottom of the vehicle and adds just a few pounds: so the impact to performance is minimal. I used a Moroso remote battery wiring kit on the Mule and was very happy with it: had everything I needed to wire the battery to the car/starter.

In general, this modification is almost always required if you are after Pro Touring-level performance—and it also helps to improve the underhood appearance.

This is what you want: the car battery solidly mounted in the rear area of the vehicle with a 2-gauge cable transferring the power to the car. Notice the Moroso quick-disconnect (pn74103) on the negative post: this allows me to quickly disconnect the battery for storage.

Battery Choices

I have run the gel-style battery (offered by Optima) instead of a lead-acid battery in most of my cars for a few key reasons. First, a lead acid battery has liquid electrolyte acid in it, and since I drive the heck out of my cars, I really don't want to get into a situation where that electrolyte would want to leak out.

The gel in the Optima battery is sandwiched into rolls of material, so it isn't going anywhere no matter what I do with the car. Also, there are many battery mounts built specifically for the Optima battery, like the DSE unit being used on the Mule, so it is easy to install one remotely in your vehicle.

Parts Used	*PNs*
DSE Remote Battery Mount	*120101*
Moroso Battery Terminal Disconnect Switch	*74103*
Moroso Battery Cable and Kit	*74055*

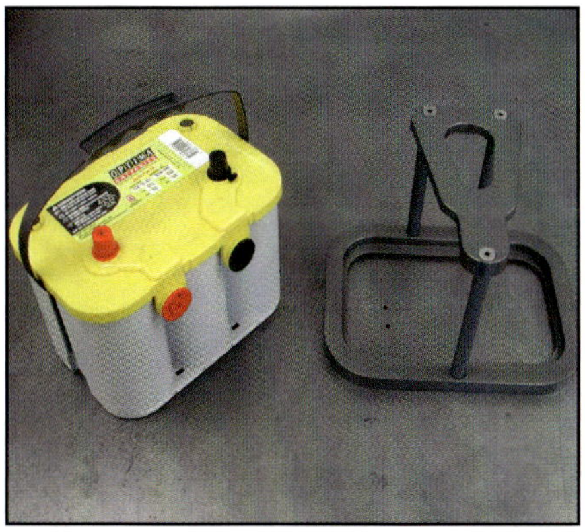

I like to use a yellow-top, deep-cycle Optima battery as it seems to handle the abuse of sitting for lengthy times, then experiencing heavy draws (from cooling fans, etc) and hard charging. Also, the deep-cycle Optima still fits in the DSE mount, so to me there is no downside to using them.

The cool thing about the DSE battery mount is how the Optima battery nests into it firmly. I used to build my own mounts, as the commercially available ones seemed horribly flimsy, but not anymore. The DSE mount was anodized to match the other aluminum parts on the Mule.

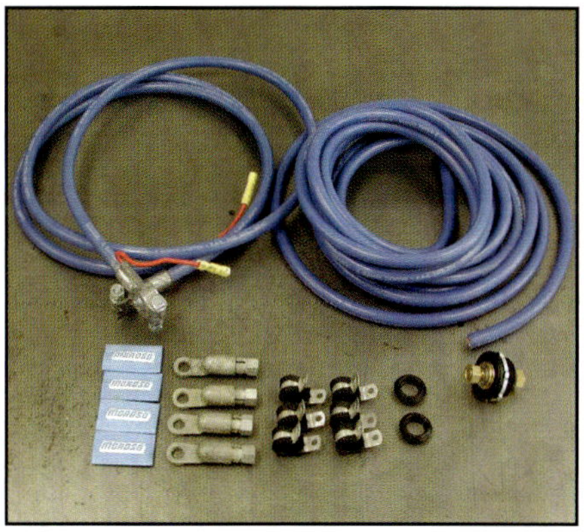

The Moroso remote battery wiring kit (pn 74055) comes with the necessary wire, connectors, heat shrink tubing and bulkhead fitting you would need to wire a remote battery into your car. It took me about a day to get this all installed.

Dashboard Tuneup

Installing the Gauges and A/C

I like to build my cars with a clean looking dash loaded with modern gauges and controls. While many years ago I used to build my own dash inserts, that is no longer the case. Detroit Speed & Engineering (DSE) now has dash inserts available for Camaros/Chevelles/etc. You will need to create holes for the gauges and controls, which is what I am doing here, but otherwise, the DSE dash inserts just bolt into the dash opening. As a caution, I always test-fit everything before sending it out for paint or powdercoating: there is a lot of 'build variation' in 60s era cars, so it's just good practice to make sure the part goes into position and the fastener holes all line up.

I use large hole saw drills (5 inch diameter for speedo/tach, 2 inch for everything else) to create the openings for my gauges in the Detroit Speed Dash Insert. While I am using my Bridgeport mill to drill these holes, I have used a table-top drill press for many years to drill these holes.

Parts Used	**PNs**
Detroit Speed & Engineering Dash Insert	*DSE010401V*
Autometer Gauges	
Speedo	*4490*
Tach	*4498*
Water Temp	*4337*
Oil Temp	*4347*
Voltmeter	*4391*
Fuel Gauge	*4314*
Oil Pressure	*4327*

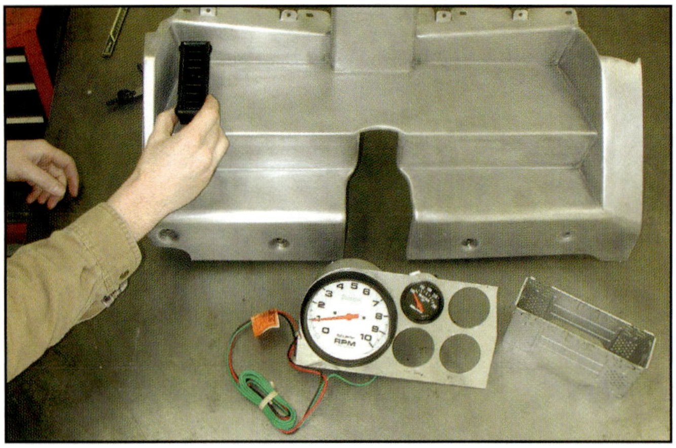

I start my layout process by mocking up the various gauges, vents and other components (like the stereo and A/C controls) on the dash insert. It is a good idea to test fit the dash insert in the dashboard to make sure everything goes together cleanly (big surprise: there was some suspect quality on many of the 60s era cars!).

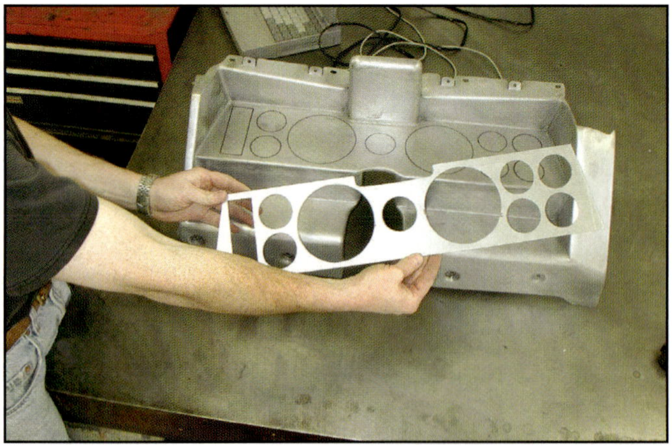

Then I create an 'insert' template using a scrap piece of sheetmetal that nests tightly into the dash insert opening.

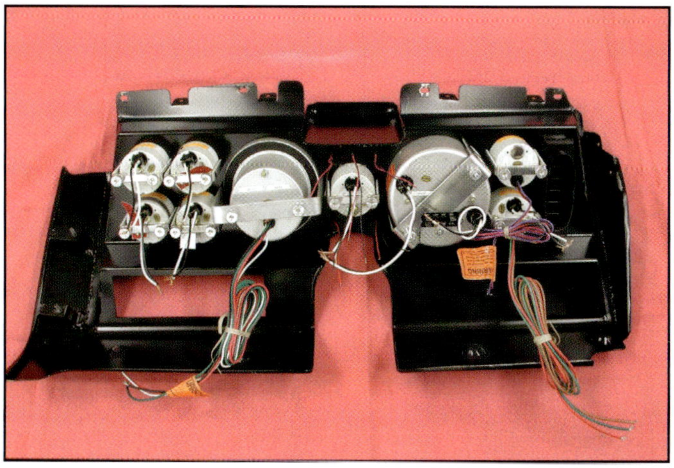

Once I have drilled all the holes and mocked everything up, I disassemble it and have the dash insert powdercoated (in this case, semi-gloss black). Then, I final assemble the gauges into the dash insert and start wiring it all into a single wire bundle that connects with a single OE automotive-grade connector.

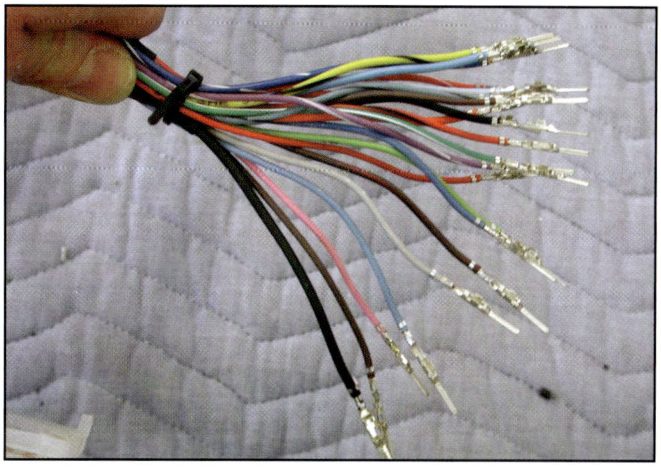

This dash was wired by the late Mike Brown, of Wires & Pliers fame. He was a talented guy and very detailed. He used different colored/striped wire for each gauge power/ground, then 'pinned out' each wire so they would plug into a 36-pin connector.

Here are the connectors being pushed into the 36-pin connector.

To me, the final-assembled dash insert setup is truly impressive with the gauges loaded into it and the wiring all running into a single wiring connector. The dash insert was then wrapped in bubble wrap and set aside until the vehicle is ready for final assembly.

The final step in the dash prep was installing the Vintage Air (I showed sealing off the holes in the firewall back on page 39 with a DSE blockoff plate) Evaporator kit behind the glove box. I like to weld ½ inch square, 0.045 inch wall thickness mild steel tubing (arrow) to create the structure that holds the Evap kit in place. I build the structure so the Evap kit can be removed and installed easily…

Gas Tank Tuneup

Building a High-G Tank

Building a 60's era musclecar to accelerate, corner and stop like a supercar requires many changes...some you don't really think about in the first dreams. One of those issues is how to get fuel out of the gas tank on a consistent basis while the vehicle is doing its thing. Seems simple, but really it isn't. As a result of me figuring out how to improve the dynamic capability of my cars, my gas tanks have gotten increasingly complex. The Mule gas tank was the pinnacle of this complexity.

Today you can buy a fabricated stainless steel gas tank for many of the popular musclecars with many of the features I am adding here (I suggest you check out Rick's Gas Tanks or Rock Valley), but back when I was doing the Mule, there was no option other than to build one myself. This chapter will show how I modified a stock gas tank to provide additional clearance to package a full-length 3 inch diameter tubing exhaust system, multiple fuel pickups and also an area to mount a large fuel pump and controller (in this case some Aeromotive product).

If you are building a Camaro/Chevelle/etc (traditional Pro Touring car), you can probably buy one of these fabricated units. If you are doing a rare Pro Touring interpretation, this chapter is for you. Without these gas tank mods, your car will more than likely need to be run with a full tank of fuel when being driven hard...and still have issues with fuel starvation in high-g loading situations.

But if you do these updates, you will be able to confidently enjoy your Pro Touring car in all situations.

The Mule gas tank was not the first stock gas tank I have cut up, but it got the most rework of any tank I have modified. Notice the safety glasses, hearing protection and gloves—yes, I always use these as I'd like to be able to see, hear and feel stuff for a long time! My suggestion is to use them and be safe: this is serious stuff we are doing.

This is the Mule gas tank after I had cut the front corners off and plated over them: I cut the corners off to provide the room needed to package the 3 inch diameter exhaust tubing between the mini-tubbed wheelwells and the gas tank. I have marked the area in the center (arrow) where I will be building an inset 'box' to mount the Aeromotive fuel pump/filter/controller assembly (otherwise it would 'crash' into the rearend suspension).

Flush it First

Whether I am working on a new or used gas tank, I always have the tank professionally flushed before performing any work on it. These facilities know how to properly dispose of the flushed material and I am then sure the tank is free of any flammable materials. If you have any reservations about working on a gas tank (and you should), find a qualified fabricator to perform the work for you.

Parts Used	*PNs*
Aeromotive Fuel Pump	*11101*
Aeromotive Fuel Filter	*12316*
Aeromotive Fuel Pump Controller	*16306*

Once I have completed the welding on any tank, I check my work to make sure there are no leaks by filling the tank with water and seeing if anything comes out. This is a low-tech test that works quite well! Either its dry and I move on or its wet and I get to keep welding.

To install the multiple fuel pickups, I cut an opening for a fuel cell access plate anchor & handmade cover. In this photo, I am scribing the outside shape of the anchor plate onto the top of the gas tank. This ring will be the bottom of a plate sandwich (with the top piece being a 1/8 inch thick stainless steel plate that I cut.

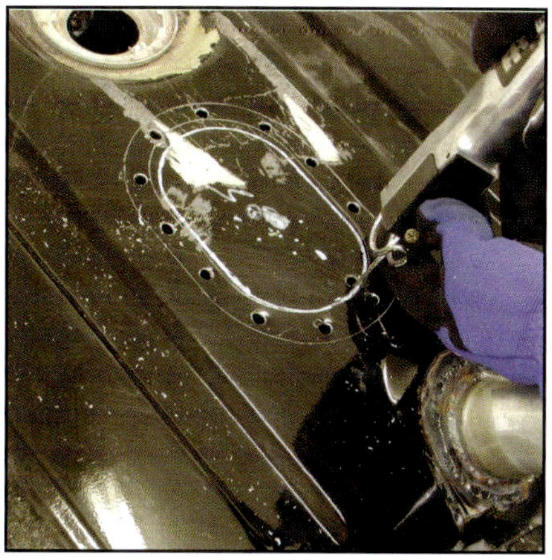

As you can see here, I have scribed the overall shape, drilled access holes for the 12 bolts the plate uses to seal off this area and am cutting out the inside opening (using a drilled hole as my starting point).

To reduce the jagged edges and blade-like nature of sheetmetal that has been cut, I dress the edges with a 2 inch diameter sanding roll. Any debris from these fabrication steps is vacuumed out and then the inside of the tank is wiped down with a solvent soaked rag before it is all blown out with filtered compressed air.

A 'step drill' (one of the greatest inventions ever!) is used to drill a hole for a -8 AN bulkhead fitting that will be installed here for the fuel line and connect the tank to the fuel pump/filter combo. A -6 AN return line goes into the passenger side, top of the tank. Again, the inside of the tank is vacuumed/wiped/blown out.

The Mule was intended to corner at high-g's, so I installed separate fuel pickups at all four corners of the tank (this is a closeup of one). These pickups are from Holley, (pn 12-951) and require a stud be welded to the bottom of the tank to mount them solidly in place (which I have done).

I then connected the four hoses coming from the fuel pickups to a common distribution block. The 'distribution block' I used was a piece of extruded aluminum fuel rail that I drilled and threaded for brass hose barb fittings on the side and an AN fitting on the end. Nice photo of the inside of the tank, huh?

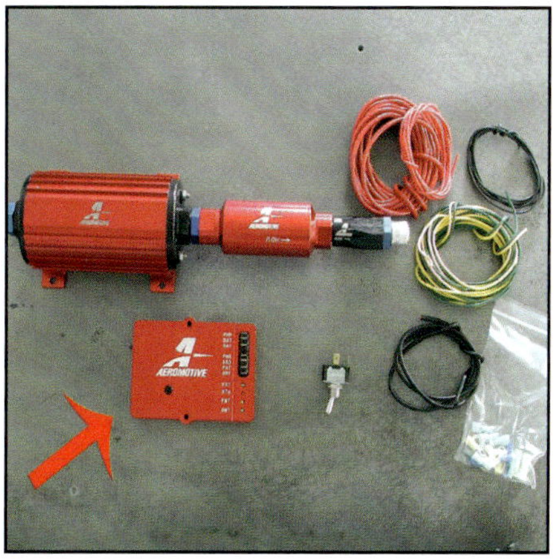

Mounts for the Aeromotive fuel pump and fuel filter were welded to the stepped plate I created on the gas tank. The Aeromotive pump was quite advanced for its day because it was a 'stepped' pump—instead of running wide-open all the time, it had its own controller (arrow) that could be wired into the throttle pedal.

One constant problem with the early Camaro fuel tanks is their filler necks leaking fuel during hard acceleration or cornering. To fix that, I weld a mid-80s fuel filler neck (with its excellent no-spill valving) onto the stock tank. This uses a screw-in fuel filler to further seal the opening.

I then sent the gas tank out for semi gloss black powdercoating on the outside and Renu gas tank coating on the inside. Then, the fuel pickup components were reassembled inside the tank and the fuel pump assembly mounted to the front of the tank before it was loaded into the stock location on the Mule to check its fit. With everything looking good, the tank was removed and set aside until the final build time.

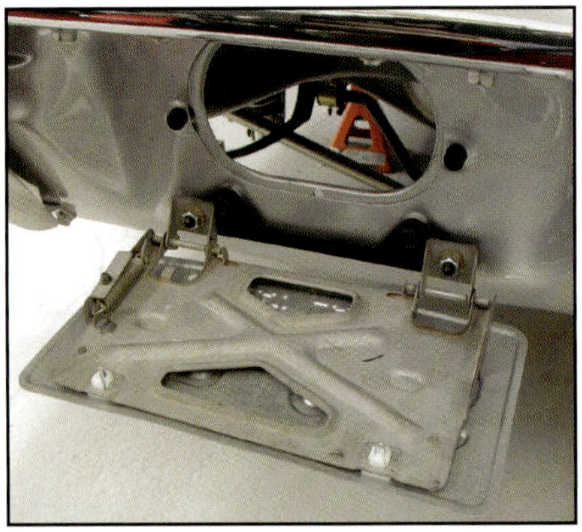

The first gen Camaro gas tanks are filled behind a flip down license plate frame—unfortunately, the original design was apt to flip down at the most inopportune time. I use the '99 Caprice license plate holder. It has a positive détente that holds it tightly in place whether it's up or down. It bolts into place using a simple adapter bracket and looks like stock.

Thrasher Gas Tank

This is the Thrasher gas tank just before going in the car. Notice the bracket welded to the front for the fuel pump/filter assembly, the welded on fittings for the feed and return lines, the access hole sealed off with an backing plate/cover plate combo and the sealed off stock sending line. This was state-of-the-art stuff in its day!

The Thrasher Gas Tank

What to do with your gas tank on a leaf spring Camaro?

As I said, the Mule gas tank wasn't my first time at the rodeo: The Thrasher tank was upgraded with a similar thought process back in late 1998. I didn't have to make as many dramatic steps with the Thrasher tank because the rear suspension was still the stock-type leaf spring design—which limited me on how big I could go with the rear tires and overall performance. But that tank was an incremental step forward in my designs. And no, you're not seeing things, the Thrasher was built when the magazines were using black & white film for tech stories, so that's why the photos are b&w! I think this is good to show how I got to the Mule gas tank and also what you could do if you have a leaf spring car going together.

I had a sump created on the bottom of the gas tank for a central capture point for all the fuel. This sump was created by cutting 0.060 inch thick mild steel sheetmetal sections (the initial plate is shown mocked up on the tank) and welding them together.

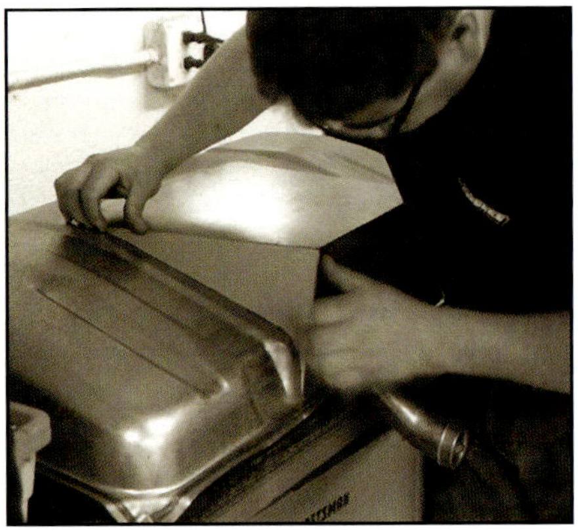

The tank was then cut out where the sump would be welded and everything final fit into position.

The sump was welded into place on the tank by a process that is called stitch-welding. This somewhat describes the process: small 'stitch' sections are welded, then the area is allowed to cool before more 'stitching' is performed. The welding area is moved around on the gas tank to prevent any one area from getting too hot.

Here I am beginning the process of drilling the holes and cutting the opening for the access area so I can install the trap-door equipped fuel capture tank. This will be closed out once everything is installed with the backing plate and a stainless steel closeout plate I built.

The trap-door fuel capture tank was purchased from ATL and was affixed to the bottom of the tank with a bracket. The fuel screen came from a '92 Camaro system to provide an added level of protection from sucking debris into the fuel pump/system.

Squeezing the fuel capture tank through the access hole, hooking up all the hoses and getting everything locked down in the tank was not a fun or easy task—but it could be accomplished and I did get it all together and working.

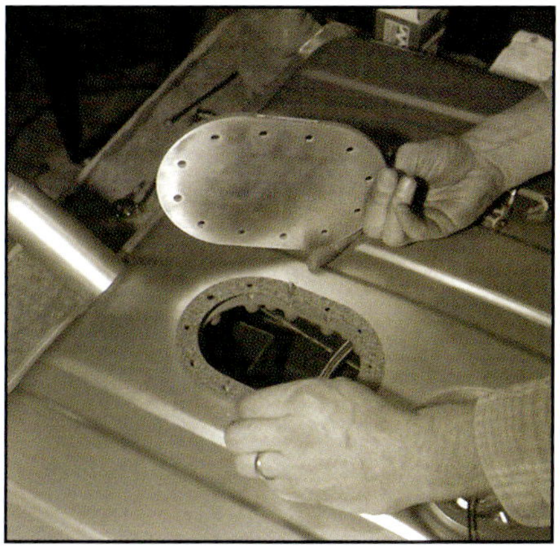

Once everything was assembled, the bracket and stainless plate were installed using gaskets above and below to seal the opening. Notice I am using two studs threaded into the backing plate to locate and hold it in position while installing the stainless plate—these are replaced with bolts that are torqued in a star-pattern sequence once all are on the plate.

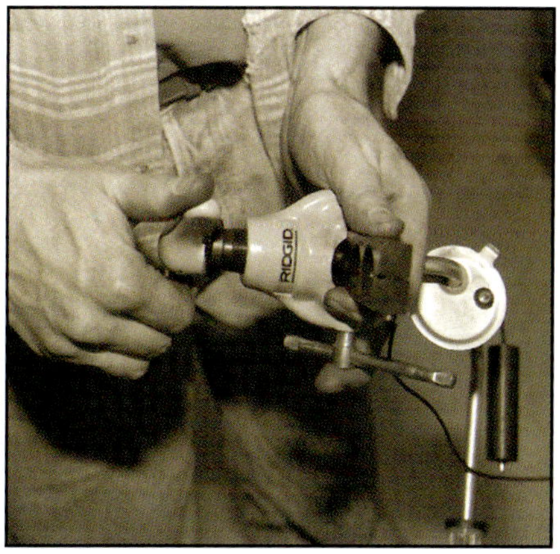

I like to use the stock fuel level senders, but didn't need the fuel pickup integrated to it, so I slid a threaded fitting on the tube and then flared the end of it so I could install a plug I knew wouldn't leak.

Here is the fuel sending unit installed on the tank with a repop fuel sending gasket from Rick's Camaros used to make sure the tank is fully sealed in this area.

The Thrasher used a Kinsler/Bosch fuel pump and Kinsler-prepped Fram filter that I mounted on the front of the tank with this plate bolted to a series of welded-on tabs. This setup worked here because the leaf spring rear suspension left enough room for the bracket to protrude forward of the tank.

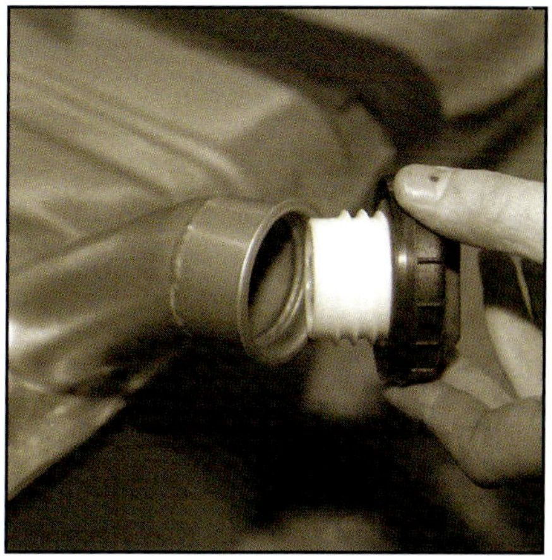

For the filler on the Thrasher gas tank, I just welded on the last few inches of an '82 Chevy Celebrity fuel filler end/threaded cap assembly. This worked, but I didn't like the way it looked. Which is why on the Mule, I welded the entire tube on the tank for a more factory-looking appearance.

In preparation for the install, I glued anti-rattle strips on the top of the tank (purchased from D & R Camaro)—these are on the cars for a reason! They reduce the rattles/squeaks/annoyance metal parts make when they rub together. Use them.

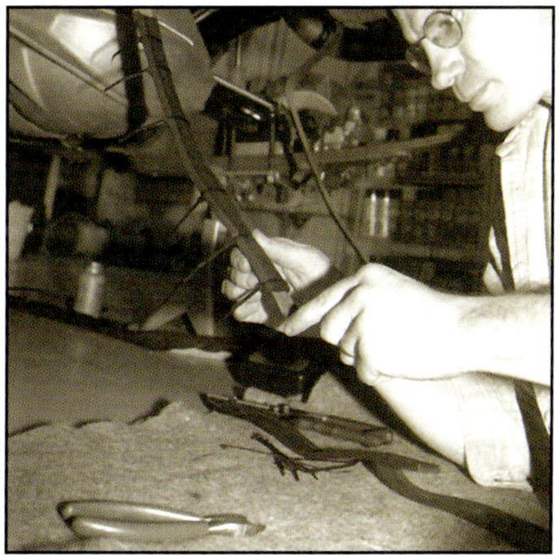

I like to use the stock fastening systems whenever possible—and the Thrasher gas tank was no exception. I glued rubber isolator strips to the stock gas tank braces and held them in position with zip ties until the glue 'set', then loaded the tank into the stock location using them as guides.

Here is the Thrasher tank installed. It worked great through everything I did with that car and it looked cool, too. I obviously didn't do the Mule tank this way, but both worked for high-g applications.

Max Flow Exhaust

Tucked tight, max flow, minimal sound: that's my ideal Pro Touring Exhaust

To me, the performance car exhaust system is often a misguided effort. Many enthusiasts equate noise with power—which too often is not a true statement... and irritating. When I am doing an exhaust system, my goals are simple: tuck it tight in the car so I don't have to look at it, make it flow the maximum amount of exhaust gas, keep the noise to the minimal decibel level, build it to last and keep it efficient in its layout. These are all difficult to achieve with one system, but I think my systems do a pretty good job of getting everything done. Here are the basics I use to make my exhaust systems great.

Starting Point

I usually start at the exhaust manifold mounting flange of the engine and work my way back to the exhaust tips. For the Mule, getting the flange of the exhaust mounting surface and long tube header flange to mate up was a major challenge. This was because I changed my cylinder head choice from the traditional small block V8 Chevy 23 degree angle cylinder head to a set of 18 degree valve angle cylinder heads. These heads were not a common choice in the market at the time, so I couldn't find an exhaust header company that made a set of bolt-on headers.

The Plan...Changes

A problem like this is not unusual when creating an exhaust system. To resolve this issue, my plan was to modify a set of traditional 23 degree cylinder head exhaust headers to fit on the 18 degree heads. I'm not sure if that was the easiest choice, but I decided to work with what I had and cut apart the Hedman headers and started the process of building my own headers.

Now I have built headers before, so I do have some experience here, but this really is one of those jobs you need to have lots of experience and confidence to make it happen quickly. Yes you can do it without previous experience, but only if you are willing to spend the time/effort and have access to the appropriate tools (see sidebar 'Tools Needed'), then I believe you could build your own headers. You just need to ask yourself if that is what you want to spend the time doing the work.

Piece it Together

Many of my exhaust systems have been built by welding together multiple 'bends' and straight sections of 3 inch diameter, 304 stainless steel tubing. These systems ended up having multiple 'butt welded' sections in them, but I have always had good durability from them.

I have usually bought these bent and straight tubing sections from outfits like Burns Stainless or SPD. I have used my TIG welder with stainless steel welding rod to fuse the pieces together into a system (after cutting the pieces to the appropriate angles and lengths). This method of building an exhaust system has taken me between 1 to 3 weeks of work to finish.

Or Buy It

On the Mule, I did things a little different than my previous exhaust systems. I contracted an outside shop, Stainless Works, in Chagrin Falls, Ohio, to create the main tubes for the vehicle. Stainless Works used their six-axis coordinate measuring arm to measure under the vehicle to 'see' where the tubing should be positioned under the vehicle, then used that information to program their CNC-bender. In fact, this system fits nicer than any system I ever built and was completed in two days! Unbelievable.

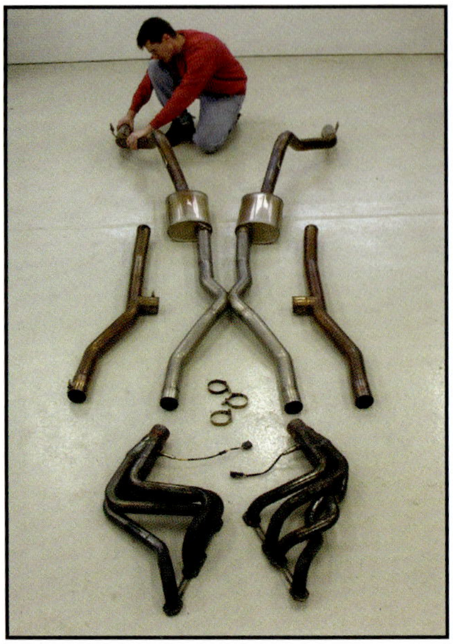

Here is the exhaust system for the Mule Camaro. As you can see, the 'H' pipe section in the middle of the exhaust was replaced with an 'X' pipe that was built by Stainless Works (and freed up an additional 11 horsepower in the process).

An additional benefit of having Stainless Works build the exhaust is they are able to bend 0.049 inch thick, 3 inch diameter stainless steel tubing, which is thinner than the 0.065 inch tubing I use to weld together. This resulted in a weight savings of between 5 and 10 pounds.

Besides building complete exhausts, Stainless Works builds their own headers, mufflers and replacement exhaust systems. They also digitize exhaust systems and replicate them in stainless steel (often requested by auto restoration shops).

Building a Pro Touring car today is probably easier than when I did the Mule as fabricated suspensions were still a very new trend, and nobody was building exhaust systems to fit around these upgrades back then. Even with professionally made exhaust systems, though, don't be surprised if you have to do a little custom work to get everything to fit, flow and sound the best it can—that's part of building your dream Pro Touring car!

The Muffler Shop

I know a lot of cars get close to being done with no plans on how to run the exhaust in them…so the builder runs the car down to the local exhaust shop to have a system slammed in. I did this with my early cars, but found these systems usually had compromises I didn't like. And that is not to get down on the exhaust shop: they are doing the best they can with a completed car. You need to decide what level of exhaust you want on your Pro Touring car, as the difference is time, money and frustration. If you don't need to fight for the last ½ inch of ground clearance, then a muffler shop system is probably OK. Just something to settle with yourself before you get to the end of the build…

O2 Sensor Placement

To maximize the life and accuracy of your O2 sensors, it is a good idea to position them angled slightly downward. This minimizes the amount of moisture that touches the tip of the sensor..

Exhaust Maintenance Tips

1. Coating After Dyno

   A good suggestion is to NOT coat your headers before you run an engine/header combination on the dyno. The extreme heat and minimal airflow experienced in the dyno has a tendency to turn the coating from a gloss to a flat finish. My suggestion is to run them bare, then have them stripped and coated afterwards.

2. Polish vs. No Polish

   If you want the stainless to look silvery after hard driving, have the stainless polished. Otherwise, expect unpolished stainless to get a gold tint to it after hard driving (I like the gold look). The polished will eventually turn gold also…

3. Anti-Seize

   If you plan on ever taking the exhaust apart/off the vehicle, I suggest you wipe anti-seize lube on the outer surface of the tube that slips into another tube. This will make it MUCH easier to separate the two tubing sections later on.

4. Band Clamps

   I like to use stainless steel band clamps as they seal the pipes tightly together. This minimizes exhaust leakage from the system--which doesn't seem like it should matter, but smelling exhaust as I drive down the road really doesn't make me feel good about my cars. They look cool, too.

Tools Needed:

Here are the tools you'll need access to and know how to use to build your own set of headers from pre-bent tubes:

- *TIG welder*
- *Band saw (or chop saw)*
- *Grinder(s)*
- *Bench Vise*
- *Various hand wrenches/tools*

Here is the issue I ran into on the 'rolled' 18 degree cylinder heads: a mismatch in the mating angle of the headers (which were designed for 23 degree heads). This forced me to cut these headers apart and rebuild them to fit the 18 degree heads—which probably wasn't such a bad thing as I usually rework headers to get them tucked up tight under a car anyway.

So here I am starting out with SBD exhaust manifold flanges and Hedman tubing bends. First issue encountered was that I had lowered the engine ½ inch to fit the T56 transmission under the stock transmission tunnel—this made some of the primary tubes hit various components…so they were reworked to fit.

As part of setting the front part of the primary tubes in position, the area where the primary tubes all come together, in what is called 'the collector', needs to be set. Work to keep the collector tucked in tight on the floorboards as this is what gets torn up during off-track forays (yeah, I know, those never happen to you either…). My rule of thumb is ½ inch between the tube and floor is acceptable—but tight!

To locate the four primary tubes together, I welded together four short sections of tube that have an inside diameter (ID) of 1 7/8 inches. This provided an excellent guide to locate my 1 7/8 inch outside diameter (OD) exhaust tubes for placement into the large collector tube.

Here is my primary tube collector guide positioned under the car on a jackstand to replicate the height I desire. I have located the guide ½ inch from the floorpan/trans mount/etc. This is tight, but if you locate everything with production-level exhaust mounts, it should provide a rattle-free system (stainless steel 'grows' a lot as it gets hot).

The SBD exhaust flanges mimic the Trick Flow square exhaust ports, so I needed 'square' tubing to fit in the openings. And here's what I came up with: a tool I cut from two sections of ½ inch aluminum plate to create a 'form' and a 'guide' I fabricated to position the tubing in.

The form shape was determined by copying the size and shape of the exhaust port holes on the SBD flange so the 1/8 inch plate could be cut/ground to shape. The form walls were then welded to the plate.

Once I had determined the length and positioning for a primary tube, I would place the tube end in the form and tighten the vise jaws down. When setting the form shape on the tube, it was REALLY difficult to get it in the exact clocked location needed—in fact, I would not have done this again as it was almost impossible to get everything lined up right.

One key is to make sure you have clearance around the header bolts to tighten them. On some of the tubes, this required the tube wall being heated up with a rosebud torch, then slightly dimpling the tube by hitting a ½ inch piece of tubing with a hammer. This made tightening the ARP 12-point header bolts a lot easier.

Here is the newly reconstructed exhaust header with the primary tubes welded in place. I like to weld the primary tubes on both sides of the exhaust flange for maximum durability. I first weld the engine side of the tube to the flange, and then let the area cool. I test fit the header on the engine, making clearance for bolts where needed. Then, I weld the backside of the tube to the flange.

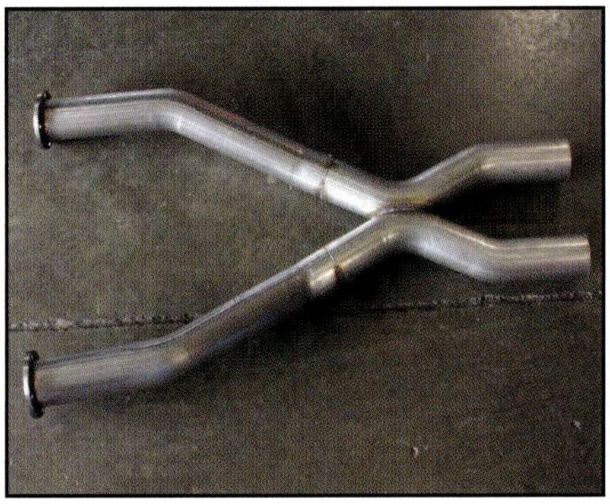

Here is the Stainless Works 'X-pipe'. These were developed in racing and have shown themselves to provide real power gains. Having a CNC-bender create these makes building them much more affordable. Notice the bolt-on flanges for the mating point to the headers. I like these because they make removing the exhaust system much easier than the slide ons (and they don't leak exhaust!).

I had the Stainless Works team position the 'X-pipe' under the driveshaft on the Mule Camaro. This kept the exhaust tucked up under the vehicle, but maximized the available space on the vehicle…there is not a lot of room under these cars!

Getting a 3 inch exhaust tube up and over the rear axle and out to the rear bumper is nearly impossible without some changes to the gas tank. As was detailed back in the Gas Tank Buildup section (page 68), I cut the corners off the front of the gas tank to allow the 3 inch tubing to pass through this area. Here's what the system looked like during construction.

Beyond building the 3 inch exhaust, just getting the pre-bent tubing section in and out of the car is a real puzzle. Suffice it to say the Stainless Works guys did a great job packaging this system (with their chambered muffler, too) in the car.

Here is a better photo showing how cutting the front corner off the gas tank opened up space for the exhaust to pass through.

I like to use Fel-Pro exhaust gaskets as they withstand most situations. One suggestion I have is to not 'rip' them out of the packaging as it distorts their shape. I suggest cutting the shrink tubing around the perimeter, pulling it off and carefully removing the gasket from its packaging.

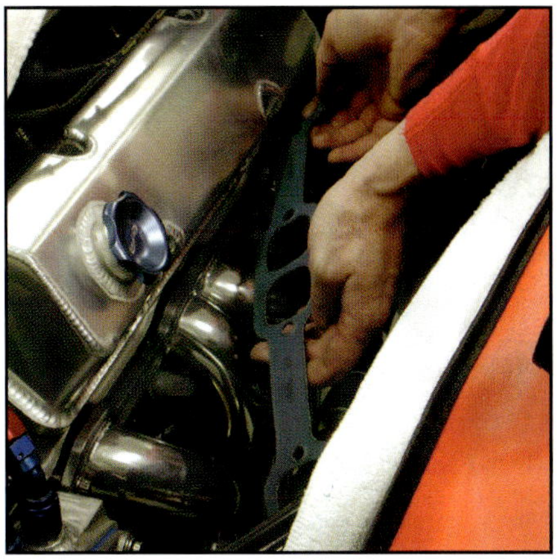

Another tip is to loosely install the header on the engine with the two outside bolts. Then drop the Fel-Pro gasket into position on these two bolts. Once the gasket is fully seated, install the remaining bolts and tighten them all down simultaneously (this usually is required or they 'bind up' on the tubes).

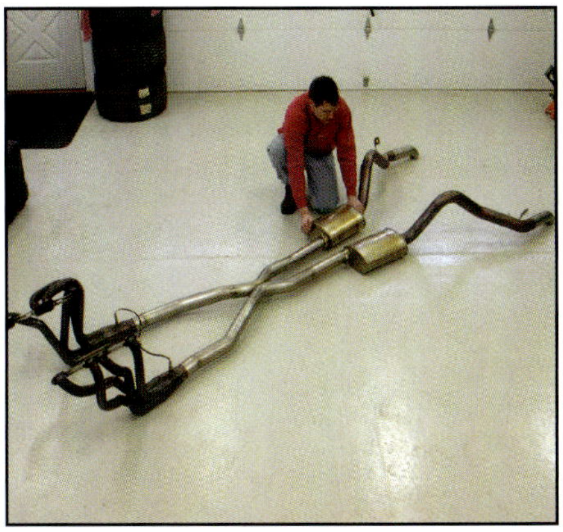

So here is the system I ended up with for the Mule. It worked great for thousands of brutal miles and continues to function quite well today—so I think it was a good design!

I like to use OE exhaust hangars where I can. On the Mule exhaust, we welded on these 3/8 inch pieces of round metal rod (arrow) in strategic locations and used factory rubber hangars/anchors to support the exhaust system. They have worked great on multiple car builds and I swear by them…

As a final tip, I suggest using these cool stainless band clamps where you are mating two tubes in a 'slip-on' joint. They seal the tubes tightly together and look great, too.

Electronics Install

Some basics for installing the ignition & engine/transmission controllers

It is really impressive how quickly the Pro Touring segment has moved forward from carburetors and points ignitions to electronic fuel injection (EFI) and multispark electronic ignitions. While everyone wants to focus on the exciting parts of going electronic, like fuel injector sizing and polished throttle bodies, I want to take a few minutes to help you avoid problems that could hurt reliability down the road.

Mounting Choices

All electronic controllers are sensitive to heat, vibration, dirt, moisture, chemicals and other elements humans aren't especially fond of either. These reasons are why *where* you mount the electronic controls is as important as *how* you mount them. I mount my controllers inside the interior, if at all possible. Usually, they end up somewhere under the dashboard, as this area is close to the engine compartment, usually has spare room (not much, but enough) and is easily accessible to get to if/when needed.

I like to keep things simple when I can, so I mount multiple electronic components to a single aluminum panel, then mount that panel in one unit under dashboard. Any electronic components should be mounted to the plate with rubber 'isolators' sandwiched between the plate and the component. This minimizes the vibration and shock being imparted to the components when the vehicle is being driven down the road and also helps the component shed heat (because there is a large air gap behind the component).

If you do this with your electronic controllers, they should provide reliable service for thousands of miles. And leave you to focus on making power, going fast and having fun with your Pro Touring fun machine.

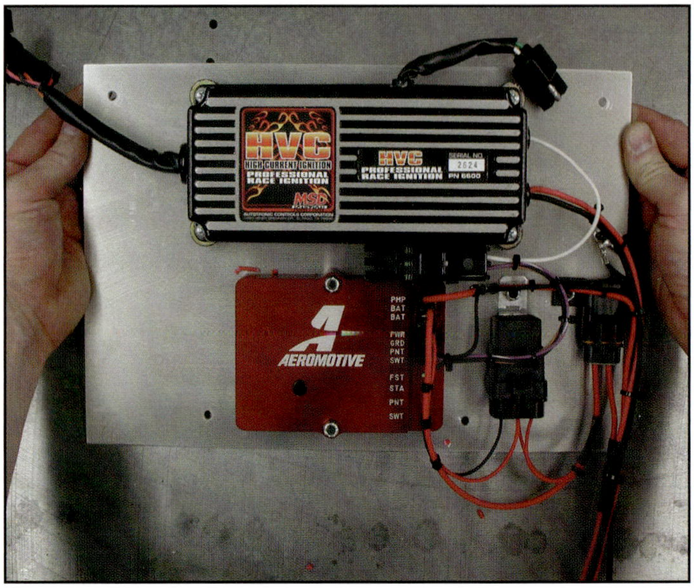

This electronic component panel was created from 1/8 inch aluminum sheetmetal to mount up under the dashboard. The large, black controller on the top is for the multiple spark ignition, while the red controller on the bottom is for the variable speed Aeromotive fuel pump. The wires connect the fuel pump controller to two relays (electronic switches). This panel installs under the dashboard on the passenger side of the body structure.

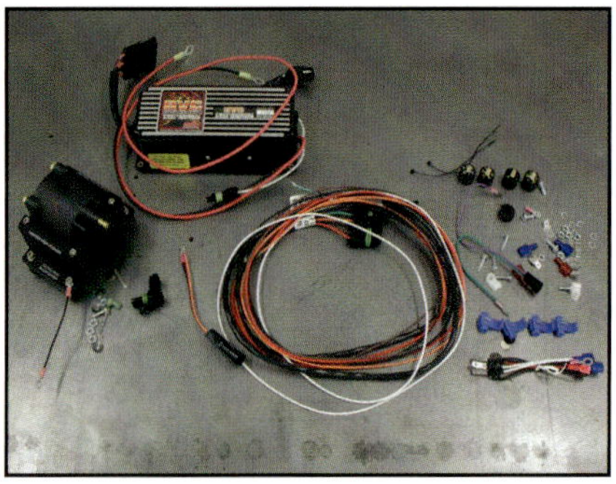

Here is what comes with a multispark ignition controller from Holley. The coil (far left) and multispark controller (top) mount in the interior and the wiring connects everything up—but I will tell you I don't use the 'pinch' electrical connectors shown with this kit. I only use OE-style connectors with pins or solder the connections. They provide the maximum reliability and after chasing enough wiring problems, that's all I use…

I thought about mounting the ignition coil on a separate aluminum plate: so I cut the piece a little bigger than the coil and drilled four holes to mount the coil to the plate. But…

I ended up deciding to drill four holes in the firewall and bolting the coil directly to the vehicle sheetmetal. My thought here was that the ignition coils are extremely reliable, so the chances of having to replace it are low—which is why I bolted it directly to the car.

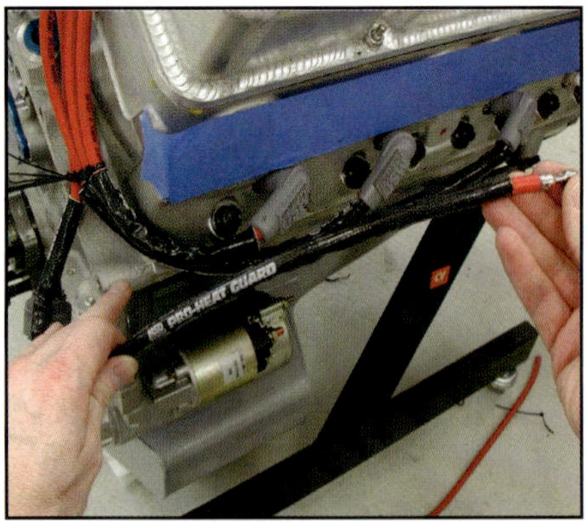

One last tip is to use high temp heat wrap, like MSD's Pro-Heat Guard sleeve, on spark plug wires that come close to the exhaust system. I don't think enthusiasts understand how destructive sustained high temperatures are to unprotected rubber & plastic components—but you figure it out when components stop working prematurely!

Shifter Rework

When building a car, there are so many things you could let slide to speed up the build process it is not even funny. Take this shifter modification. I could have just punted and forever 'reached' for the shifter every time I needed to make a gear change. Would I have liked driving the car? Probably not. Which is why it is these 'little things' that require the time and effort to make your car 'right'.

Take a look at how I found this issue, then how I solved it.

This seems like a simple issue: I install the seat I want with the T56 transmission installed and realize the shifter arm positions the handle three inches from my hand. I checked to see if there were any different shifters available (there weren't), then decided to extend the handle arm a few inches and changed the threaded end so I could run the shifter ball I like.

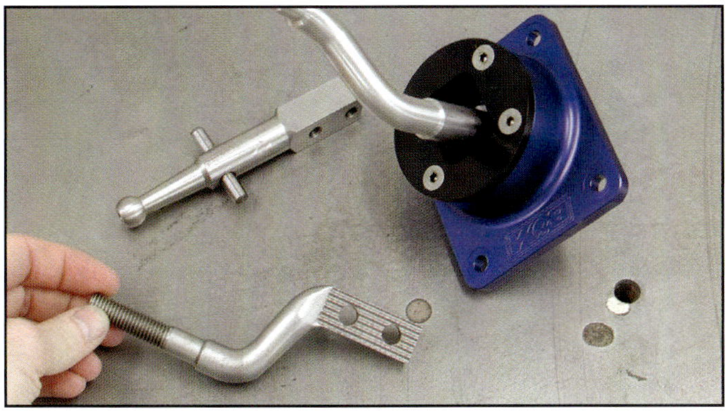

I started with a B&M short throw shifter assembly built for the T56 manual transmission that came with a single piece arm installed. I decided not to use their shifter as I thought it would be difficult to lengthen it and keep the shifter handle low in the interior.

The shifter arm I decided to use works with a 'universal' actuator arm (that is being installed here) that allowed me to try a few offset shifter arms. I settled on the shifter arm shown here (arrow).

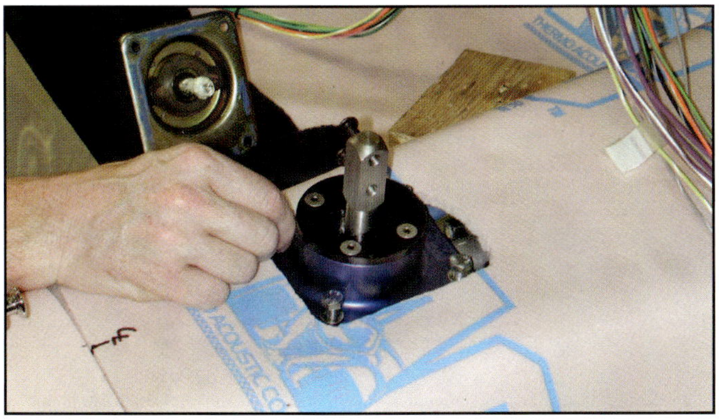

Installing the B&M Shifter base on a T56 transmission couldn't be much easier: remove the four bolts on the stock shifter, pull it out, install the B&M shifter, torque the four fasteners.

I marked the shifter before cutting it in half. The lines represent the cut line (vertical line) and the centerline of the shifter arm at that point (horizontal line). These lines will both come in handy as the modification comes together.

Using my trusty band saw, I make a straight cut down the center of my 'cut line' on the shifter arm.

Using my grinder, I beveled the edges of the shifter arm pieces to create space to for adding weld material (when everything goes back together).

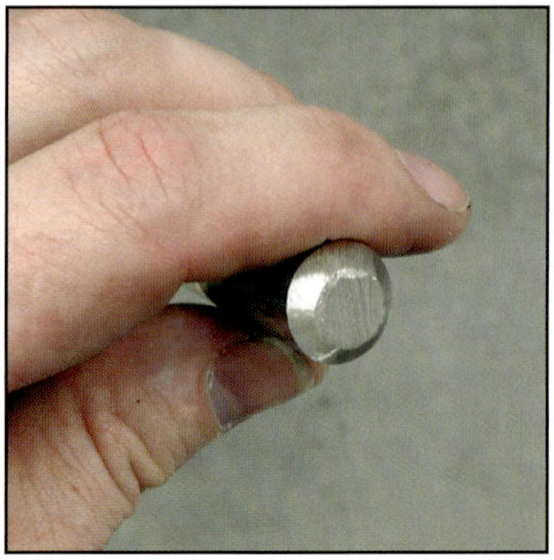

This closeup gives a good idea how much material was ground off the edge of the cut ends.

I then cut and beveled the ends on a 3 inch long, ½ inch diameter, solid aluminum piece of round bar stock and positioned it between the two pieces of the B&M shifter arm on my surface plate.

Using machining guide stock to hold everything in position, I clamped everything in place on my surface plate in preparation for applying a few TIG tack welds to check my work.

Here I am adding four tack welds (two per mating point) to the shifter so I can install it in the car for a test fit.

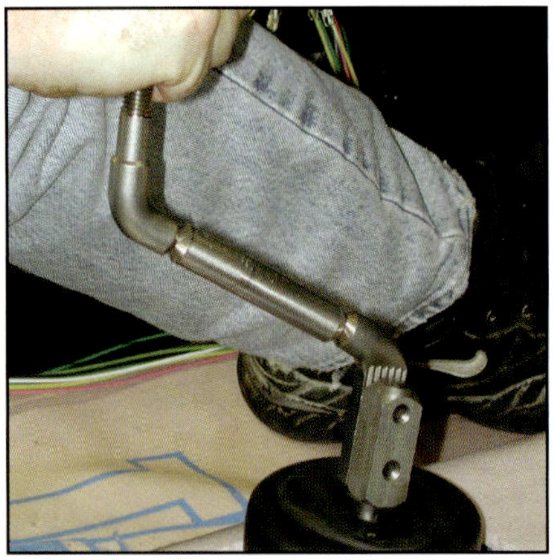

The new location turned out perfect, so the shifter was unbolted from the T56 transmission and fully welded into one piece. I then ground the mating points completely smooth and welded a new threaded end on the shifter arm so it would accept a very cool shifter ball I had found.

Here is the lengthened shifter arm after all the modifications—looks good, doesn't it?

I cut off the head on a ½ inch diameter aluminum bolt and welded the threaded portion of the bolt to the end of this shifter so I could run my favorite shifter handle (a ball).

Paint Shopping

Here's the way to get a good paint quote: A bare car, a trailer and time

I have had about 15 cars painted at paint shops over the years and am very clear on how to end up happy with a paint job. Most of my lessons have been learned the hard way, so you can either listen to what I have to say and use it, or go off and learn them yourself…but don't say I didn't warn you.

This is the only way I will talk to a paint shop about putting together a quote to paint one of my cars: with a car that has been stripped of all its paint and body filler, all the fabrication work is completed and all the components are with the car.

Strip it

The most important step you can take to end up with a great paint job performed at a competitive price is to have the vehicle stripped before you go to the paint shop. I like to have the body blasted with media, as this strips the paint and body filler but is less prone to warping body panels or otherwise damaging the sheetmetal (like sandblasting can sometimes do).

Modify it

It is very important that you have all the heavy fabrication work completed before you go to the paint shop for a quote—as it is something else they will need to be able to see and understand to say how much work is involved. A good example of this is a rollcage: painting rollcage tubes is a bitch. The entire car needs to be taped/papered off to prevent overspray and the painting itself is really tough to do, so the cost of this is often substantial.

Get Multiple Quotes

Even if I like the first quote, I have found it is very important to go out and get more quotes as I always learn something about the paint process on the car in question. Make sure to get at least three quotes and spend some time understanding why they cost what they do and which shop offers the most value to you for the quoted cost.

Verbal Changes = Cost Overruns

If you make lots of mid-stream verbal changes, expect to pay through the nose for them. Some shops plan on you doing this and make their profit via them. Others just charge for the inconvenience of you waffling on what you want. Either way, mid-paintjob changes brutalize the budget. My suggestion is to spend the time prior to the quoting process to decide what you want, clearly communicate it and then hold the shop accountable to completing that paint job.

Punting

There have been times when I had to rescue a car out of 'paint jail'. These are sad moments where I have had to admit I made a bad choice and things are not progressing on the paintwork. Its moments like these that have made me very clear on how to do the selection process properly to maximize the chances of getting a great paint job, on time and on budget.

Top 5 Paint Quote Tips

1. *Bring a car to the paint shop with no paint or body filler and all fabrication work completed on it*
2. *Tour the paint shop: if it is loaded with cars covered in 5 years of dust and the shop rep is griping about the customers, run, don't walk, to find another shop to do the work*
3. *Bring all components you wish to have painted to the shop*
4. *Have a list of work you desire to be completed: bodywork, priming, painting, colorsanding, buffing, assembling (hanging all swingmetal)*
5. *Discuss type of paint products you want on the vehicle (epoxy primer, 1-, 2- or 3-stage paint, etc)*
6. *Walk out with a quote!*

E-Coat: Not a Good Idea

Sometimes things that don't work out

Most of the 'new' products or processes I have tried over the years have worked out quite well for me, but there have also been some 'duds'. One of those 'not so much' moments has been my choice to have the Mule body and chassis components 'e-coated' before sending everything out to be painted.

Actually, I don't think the e-coating was the issue, I think it was the chemical stripping that was required before the e-coating process that was the problem. That's because it appears the stripping chemicals never stopped 'stripping' the metal—eventually leading to accelerated corrosion deep within intricately spot welded assemblies (like the Mule body or doors).

My point with listing this is that I do like to try new things and I know a small percentage are not going to work out. To me, building and integrating new ideas/ components is worth the effort and occasional failure. Now, I don't suggest you just slap on anything you come across. You need to do some research. But if you feel the product is worthy of being installed on your car after looking into it (on the web, forums, etc) I would encourage you to try it. You might be happily surprised—I know I usually am.

What is E-Coat?

E-coat is a shorthand term for 'electrocoating'. It is a process where a part is electrically grounded and immersed in paint that is charged with the opposite polarity. Because of the electrical charge, resin and pigment coat the part resulting in a uniform film across every inch of a part. The e-coat provides an excellent base for applying paint, so the OE automotive business uses it on almost every body, chassis or suspension component.

To e-coat the Mule, the sheetmetal was chemically stripped, then strapped to a fixture that would go through a production e-coat process. I had high hopes for this process and, in general, was happy with the e-coating. But in the end, this didn't work out for the Mule.

The e-coat process was impressive to watch happening as the vehicle was dipped into tanks big enough to accept an entire car body—lots and lots of tanks this size filled with chemicals. The fumes were quite strong, but the process itself is in slow-motion as it moved from tank to tank.

Here is the Mule leaving the rinse tank with the e-coat on all the surfaces. The OE-business e-coats complete bodies, but they are brand new, so they don't need to be chemically stripped. I will say, I would have a 'repop' body e-coated (since it wouldn't need to be stripped), but this company is out of business, so that is not possible.

Powerplant Buildup

A straightforward, lightweight 540hp Small-Block Chevy

After building and running practically every engine known to the Bow Tie world, you might be wondering why I'd build a plain-Jane, small-block Chevy to power the wacky conglomeration of one-off prototype parts that are the Mule. Have you ever needed to find shaft-mount aluminum rockers for a one-off splay-valve, small-block Chevy in the middle of Missouri? No? Well, I have, and they don't exist. That's why.

Through life's painful lessons, I have learned that the oh-my-gosh, hood-raising factor of an "unobtainium" engine is a certain kind of cool, but if the car is going to be really driven, nothing beats the simplicity and power potential of a small-block Chevy V8. I think it brings on the thought of a subtle cool, which is what this engine represents. It might look like a regular small-block, but this engine is loaded with its share of charm.

Take the beautifully created GM Performance Parts aluminum engine block, the lightweight Scat crank (only 47 pounds!) and the "fully CNC-machined everything" 15° Chapman cylinder heads. These and the rest of the engine parts used here have benefited from years and years of refinement on the racetracks around the country and combine to produce exciting performance with minimal weight.

The Big Decisions

I built this engine combination to 400 cubic inches as I felt it would provide the maximum power potential in the small-block architecture. I have built Gen I small-blocks of larger cubic inches, but to go much bigger than 400 ci requires expensive custom parts and modifications to the block and other components, all of which I'm steering clear of. That's not to say the Scat crank and rods aren't trick--they are--but they're also sitting on the shelves of many speed shops, ready to ship.

On the street, I feel having an engine that will rev quickly is just as important as making big power. For that reason, the reciprocating components have been selected to minimize weight without hurting durability. I really drive my cars hard, so the parts need to take serious abuse without a lot of maintenance. The crank, rods, pistons, rings, and valvetrain were combined to provide the least mass and frictional losses as possible in the system without compromising durability.

The fuel mixture will be controlled by a port-injected, throttle-body controlled electronic fuel injection (EFI) setup similar to systems I've used on previous vehicles. The preciseness of fuel injection makes the engine crisp, powerful, and repeatable in practically any driving situation I can throw at it, which is very important to me.

I have assembled engines before, but like body and paintwork, I prefer to have a specialist do the work since they develop wisdom based on experience (which usually involves breaking parts!). In this case, I'm having well-known engine builder Kurt Urban prep the parts and assemble the engine. Urban builds engines out of his KU Performance Shop. He applied his extensive knowledge about building aluminum block engines for the street-stuff I didn't know, which made me happy to have him doing the work.

As much time as I spend building my cars, you'd think that's what I really love--but it isn't. My favorite aspect of building fun cars is driving them. While this engine is light & subtly trick, it should act a lot like a "driver." The only special treatment it will require will be on cold mornings. Urban has said it will need extended warm-up time on cold mornings to ensure the aluminum block is up to temperature (the bearing clearances are tight to take into account the expansion of the block). Other than that, I plan to drive it like a rental car!

While it looks like just a small-block Chevy, this potent 400ci engine is about 170 pounds lighter than a production small-block. It'll make a tad over 540 hp, run on pump gas, and be docile enough to drive to work.

CNC Beginnings

Back when the Mule was built, computer numerical controlled (CNC) cylinder heads were still a pretty trick deal...now everyone has them. When I chose the Chapman heads, I was really pushing the limit and was very happy with the results.

The Chapman 15° heads started as raw GM 18° castings (originally designed for the top ranks of NASCAR stock car racing, before they converted to the SB2.2 design years back) with the head mating surface faced to locate the head in the CNC machine. From there, the intake and exhaust ports and the combustion chambers were machined to their final shape in the CNC machine. Many of Chapman's products, like these 15° heads, are now available through Scoggin-Dickey Parts Company in Lubbock, TX. I got these heads because I felt they would make excellent power due to the fully CNC-machined ports and chambers and because they could take the abuse of hard street driving.

Chapman creates the 15° valve angle by "rolling" the head 3° before machining it in the CNC. This process raises the intake ports up about .150 inch compared to a stock head, promoting airflow both into the intake port and past the intake valve. To mate the intake and heads, a set of spacers are needed to get proper port alignment of the GM 18° lightweight manifold.

If you've never seen a five- or six-axis, CNC-machining center chiseling away on a cylinder head, you don't know the magic of CNC cylinder heads. As far as I'm concerned, this is some of the most awe-inspiring racing technology available--the technology will improve the performance of practically any engine. The CNC machine can be programmed to do multiple machining processes, which is what Chapman does (and used to have to do by hand, with repeatability only in the direct hands of the porter!)

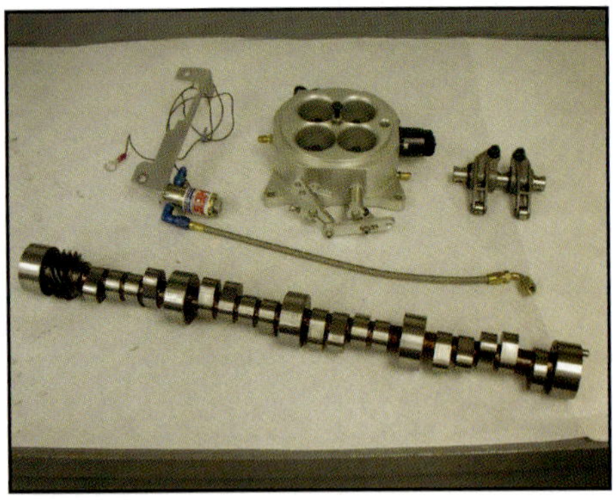

The 'Street' Package

When building this engine, I tried a few different camshaft and rocker ratio combinations when we were on the dyno. My first goal was to make over 600 hp—which we did with what I would call the 'big cam, high ratio 1.7:1 rockers' combo. I ordered a 'small cam' at the same time, along with a set of 'low ratio' 1.55:1 rockers. Engine builder Kurt Urban and EFI guru Dave Henninger tested these parts in various combinations to develop an ideal for my application.

NOTE: All dyno testing was performed with the 400ci, EFI, all-aluminum small-block Chevy engine running through a complete Stainless Works exhaust system (yes, we ran it with the mufflers and headers that we'll run in the car), 94-octane Sunoco pump gas running about 12.9 to 13.0:1 air/fuel ratio, non-detergent 30W Valvoline oil at 50 psi and 190° F, water at 180°w F. The figures shown here are corrected to SAE standards.

Dyno Results

RPM	Big Cam 1.7:1 Rockers		Small Cam 1.7:1 Rockers		Small Cam 1.55:1 Rockers	
	Torque	*HP*	*Torque*	*HP*	*Torque*	*HP*
3100					438	250
3200					443	262
3300					450	274
3400					459	288
3500	453	302	453	302	463	300
3600	456	312	456	312	466	311
3700	460	324	460	324	467	320
3800	463	335	463	335	467	329
3900	464	345	464	345	467	338
4000	465	354	465	354	467	347
4100	465	363	465	363	468	357
4200	470	376	470	376	473	369
4300	476	390	476	390	477	381
4400	484	406	484	406	482	395
4500	490	420	499	428	491	412
4600	498	436	505	443	506	443
4700	506	453	511	457	501	430
4800	511	467	513	469	506	443
4900	513	479	514	480	510	456
5000	514	489	516	491	510	467
5100	517	502	516	501	511	477
5200	519	514	517	512	509	485
5300	523	527	518	522	509	495
5400	525	540	517	532	508	503
5500	526	551	516	540	506	511
5600	525	560	512	546	504	519
5700	525	569	509	552	502	525
5800	519	574	502	555	499	532
5900	516	580	496	557	495	537
6000	514	588	488	558	488	539
6100	510	592	483	561	481	541
6200	505	597	477	563	473	541
6300	500	599	470	564		
6400	495	603	461	562		
6500	491	608	454	562		
6600	485	609				
6700	477	608				
6800	471	610				

A Day On The Dyno

Dyno time is usually sold by the day or hour, so here's a breakdown of the Mule engine's day on the dyno.

7:00 a.m.	Roll engine into dyno room. Connect flywheel, water, oil, electric, etc.,
8:30 a.m.	Spin engine to determine if all connections are working properly. Build fluid lines and adapt various systems to dyno to get everything working right. Load initial calibration into EFI controller.
9:30 a.m.	Fire engine to check for leaks, etc. Fix leaks. Refire. Warm up engine.
10:15 a.m.	Make initial power pull. Begin tuning calibration by stepping to rpm, holding it there, and setting air/fuel ratio and ignition timing based on O2 signal, manifold pressures, and power output.
Noon	Time for a full 'sweep test' pull! Continue refining air/fuel ratio, ignition timing and retesting to determine max power production.
2:00 p.m.	Swap parts (throttle-body spacer: lost 5 hp; low dB mufflers: down 16 hp, but 6 dB less, etc.) and retest with 'sweep test'.
4:00 p.m.	Whew. Remove engine in one piece from dyno. Package up to take home and install in Mule.

Camshaft Specs

Camshaft Specs	Big Cam:	Small Cam:
Advertised Duration @ 0.015 inch lift [degrees]	302/310	235/244
Operating Range, rpm	3500 – 7000	3000 – 6500
Opening/Closing Point [degrees]		
Intake	19/53	11/45
Exhaust	67/13	59/5
Maximum Lift, Int/Exh [degrees	107/107	117/117
Duration, Int/Exh [degrees]	236/244	252/260

TOP 5 ENGINE-BUILDING "D'OHS!" (OR "DOUGH")

1. Starting without enough dough ($).
2. Making compromises due to slow part delivery or lack of dough ($).
3. Taking more time then expected (and time is dough—$).
4. Changing the plan midstream and spending lots of extra dough ($!).
5. Always costing more dough ($) than you expected.

Thought vs Reality

1

THOUGHT: Engine's like dyno testing

REALITY: Many just break to avoid the work.

2

THOUGHT: You can dial it in on the dyno

REALITY: You need to prep everything before getting to the dyno-ignition wires, fluid hoses, wiring, etc., because DYNO TIME IS DYNO TIME. You pay whatever per hour whether the engine is running or not.

3

THOUGHT: Dyno testing is fun and exciting

REALITY: It's about as stressful as can be. The joy is in completing the session with good power and not carrying the engine away in a soggy brown paper bag.

If you're wondering why you would run an aluminum block, here's the reason--they're light! It weighs only 92 pounds, while a GM Bow Tie cast-iron block is 202 pounds. The aluminum block comes with doweled, four-bolt main caps and is meticulously machined for GM by Schwartz Machine.

One of the details that Kurt Urban is very clear about is using a grinder to remove any casting flash or parting lines on the inside surfaces of the block. Aluminum expands at twice the rate of iron and is less rigid, so the metal in the block is experiencing a lot of forces. Any sharp points in the metal have a higher chance of being a "stress riser" or a point where a crack would begin. These stress risers are more critical here than with an iron block. He also doesn't like to have sharp edges that might cut his hand when he's building the engine!

Another step to maintaining oil control is preventing the oil from pooling in the valve cover (and being sucked out the breathers). To do this, Urban recommends threading the two feed holes to the lifter oil gallery and installing these .060-inch restrictors. The center hole is the feed from the oil pump.

The only risk in doing this is running the valvesprings short of lubricating and cooling oil, but I'm running special valve covers from Billet Fabrication that spray a light mist of oil directly on the valvesprings, so there are no worries. By the way, I chose to not run one of these restrictors in my PHR 427 ci stroker—which is why it blew up at Sebring in the One Lap! I learned my lesson there....

Since the aluminum block "moves around" and "grows" a lot as with an increase in temperature, the tolerances like those for the main bearings are tightened to reduce the amount of oil flopping off the crank.

Urban drills additional oil feed holes, 0.125 inch in diameter, (arrow) on each side of the main bearing to the road race motors he builds to insure there is sufficient oil to the rod bearings. The location of the hole on these bearings needed to be lower than what he would normally put on his cast-iron bearings as the oil slot in the block didn't go as far along the side of the bearing.

For an aluminum block, Urban sets the main bearing clearances at .0018 inch, versus .0025 inch on an iron block (this is on everything but the rear main). On the rear, he runs .0025 inch versus .0030 inch with iron. Urban told me, "Stielow, you'll shift this thing at 7,000 rpm, and I want a little oil on the thrust bearing to prevent a problem." We've brought him back problems before! Here's a tech tip: Urban always line hones and measures the mains with the oil pump bolted to the main cap. "It's amazing how much the torque of the oil pump stud will pull the main out of round," he says, "So I always bolt up a dummy oil pump before doing that machine work (arrow)."

Checking the crank fitment in the engine and establishing the bearing tolerances requires it be installed and measured. To duplicate the stresses on the block when it was initially machined, a very specific torquing procedure is followed. Even the thread lubricant provided by GM must be used. Urban suggests following this exactly: "GM spends unbelievable time and money to get the most from this equipment--the least we can do is follow their instructions on torquing." Notice how Urban has written notes to himself on the block regarding the tolerances he saw on the mains during his initial mock-up of the parts in the block.

Even with a lightweight crank, the counterweights on the Scat crank hit the extra thick main webs (the aluminum has less strength than the iron, so the webbing is needed to be thicker). Urban solved this by having the crank counterweights machined by John Kominski to remove about .120 inch at the very edge of the main web-face. This created a 5-degree angle on the counterweights. Kominski then balanced the rotating assembly. To finish off the clearancing, the webs on the block were slightly reworked with the grinder. When completed, there was .060-inch clearance between the crank throw and main webs of the block.

There are many slick features on the GM aluminum block, but one of the most visible components are these threaded, anodized, and O ringed freeze plugs located all over the block. They are tightened in place with large diameter Allens and seal the engine tightly.

Roller cams must be bushed very carefully to keep them from "walking" forward and back excessively in the block. They move around because roller lifters don't rotate around their centerline like flat-tappet lifters, which hold the cam in place. For roller-cammed engines, Urban spends considerable time choosing and machining a cam button to hold the cam between .012- and .020-inch clearance. The o-ringed rear cam cover is held in place with countersunk Allen bolts (and engraved with the GM Racing logo).

Here's another quick tip: Never install a part straight out of the box. Even if the part looks super clean, random pieces of packing materials and cardboard particles can clog an oil passage faster than you know. Whether they're straight out of the box or just back from the machine shop, Urban lets the solution of a solvent tank pour through the passages of parts for a long time, then scrubs everything thoroughly to make sure any trash is washed out.

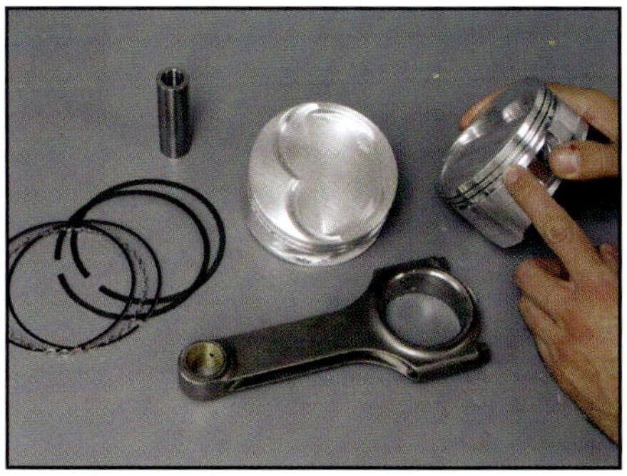

I have used Arias pistons, Speed-Pro rings, and Scat rods in the past and must say they have never let me down. On the pistons, notice they are a lightweight design but the ring package doesn't need a spacer ring to support the oil control ring--a common feature when the wrist pin bore intrudes into the oil ring land. Not having this reduces piston weight and simplifies the ring package (which is made up of .049-inch/.049-inch/3mm rings from Speed-Pro).

To establish the tolerances for the big end of the Scat rods, Urban assembles them with the bearing halves in them and the threads of the ARP rod bolts wiped with assembly lube.

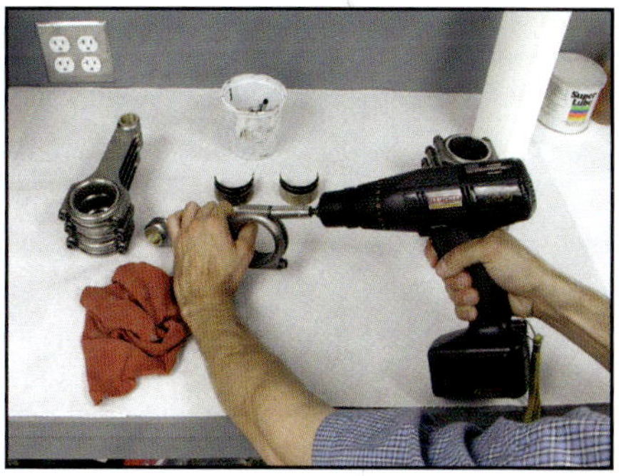

All engine builders now use electric wrenches or drills to reduce the strain on their bodies, and Urban is no different. He was using these years ago to get the fasteners hand tight, then he would torque the fasteners to 20 ft-lbs and then adds another 50 to 70° of bolt rotation until he gets to the proper 'bolt stretch'.

The bearing halves are mixed and matched via mocking them up on the engine until the difference between the rod big end bearing opening and journal size equaled 0.0025 inches.

I had bought a set of Arias pistons for a different engine combination a while ago, and found they would work for this 15-degree Chapman heads if the domes were machined down (right) 0.080 inch. The machined areas were sanded smooth (left) before being installed.

The Arias pistons were then measured so they would fit properly in the bores, then assembled on the ends of the Scat rods (after being cleaned and oiled up) and loaded into the bores.

The GM BowTie block needs to have Allen plugs installed in the main oil galley with thread sealer. These are in the block because they are used as access points for drilling the priority main oiling passages.

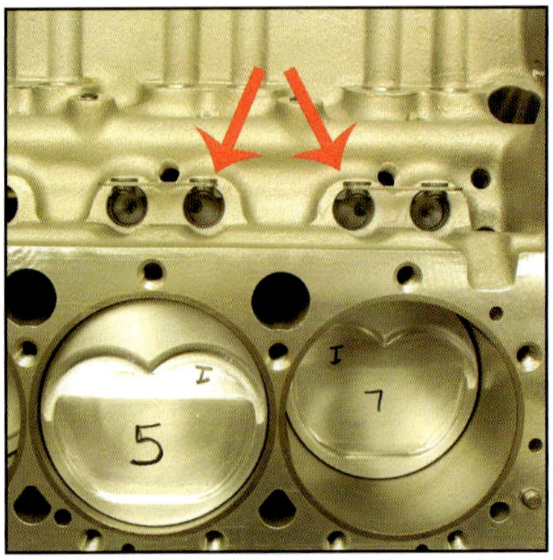

To fit the pushrods around the big intake ports, the Crower intake lifters are 'offset'. Note how Urban writes what hole (cylinder bore) the pistons are being assembled for. These 'notes' are key to 'remembering' where to put everything during the final build process.

Here is how the lifters are situated in the engine.

The best way to check the piston-to-valve clearances is to place small pieces of clay on the top of the pistons (Urban uses a 'T' shape, like shown) under the intake valve, torque the respective cylinder head on with one set of valvetrain/head gaskets/etc on the engine and cycle the engine 360 degrees. Then remove the parts and measure the depth of the clay that is remaining. The clay should be at least 0.120 inch on an aluminum engine (0.100 inch on a cast iron block engine).

Here is the combustion chamber on the Chapman 15 degree cylinder heads after Urban had polished them smooth (to remove the CNC ridges). Notice how 'flat' the chamber looks—that's the advantage of running a cylinder head with a small valve angle.

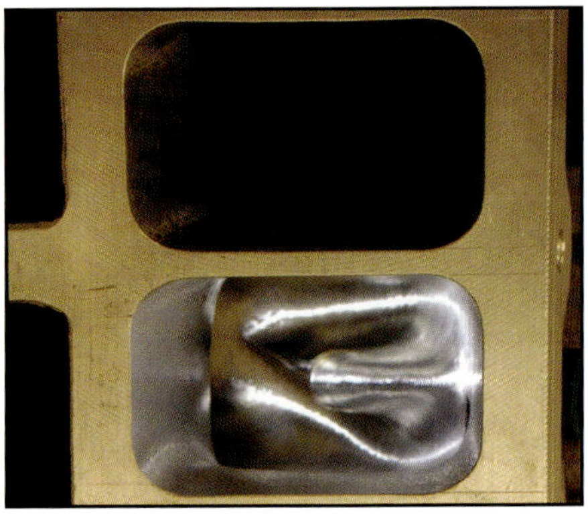

Here is a shot into the very refined intake port on the Chapman head. Notice how the valve guide is sculpted to maintain maximum airflow (top) and the port helps to create swirl into the cylinder bore. Trick stuff…

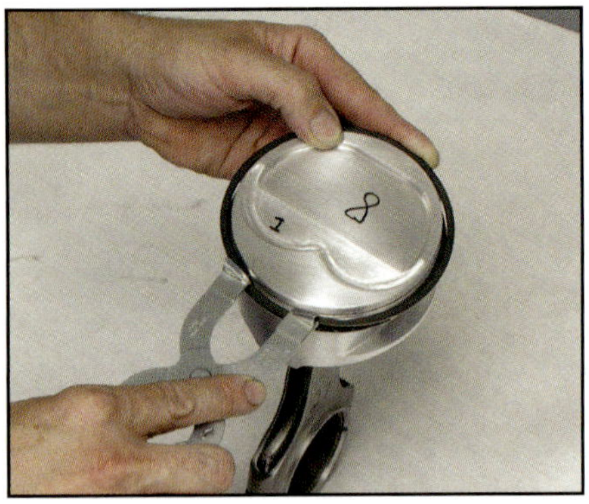

Next, the pistons were final assembled with the Speed Pro Rings

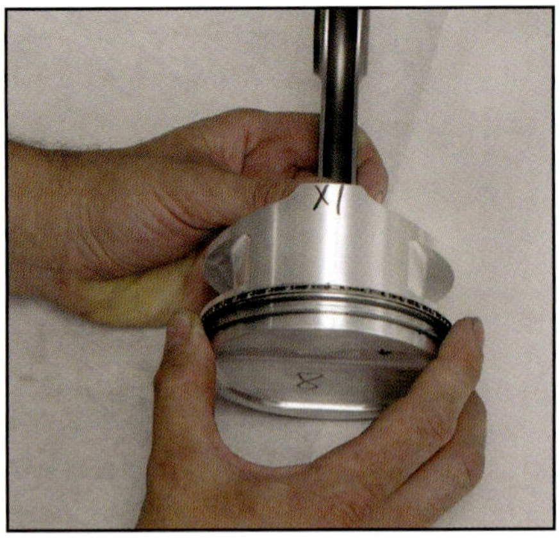

Urban has spent many years building engines and one facet of knowledge he has gained has been on where to place the gaps of the rings on the pistons--a process known as "timing the rings." You see the Top Fuel guys doing this right before they stick a piston in a bore on TV. For this engine, Urban placed the ring gaps for the top ring and the oil expander gap facing the non-thrust side of the piston, and the second ring gap and the oil rails are clocked on the thrust side.

Urban spends a lot of time honing the cylinders to achieve a very specific diameter and surface and then washes the block thoroughly. He grinds the ends of the rings to get the piston rings gapped for each cylinder (this engine has gaps of .018 inch on the top, .022 inch on the second), installs the rods, uses a ring spreader to install the rings on the pistons, lubes the piston-thrust faces with non-detergent 30-weight motor oil and slides them in. Notice on each piston he has written the cylinder number and thrust face.

Urban is a big believer in using lubricants designed for specific applications. Assembly lube is liberally poured on the bearing surfaces and wiped across the face, and ARP moly thread lube is slathered on the rod bolts. Torque readings can easily be fooled by using a different lubricant than the component manufacturer recommends so it's important to use the same lube the manufacturer uses--not doing this could lead to inaccurate torque readings and the disastrous loosening of fasteners after the engine is running.

Once the bearing tolerances were attained through mocking up the parts, everything was cleaned once more, oiled up thoroughly and final assembled. Notice how the oil pump is final installed on the engine at this point.

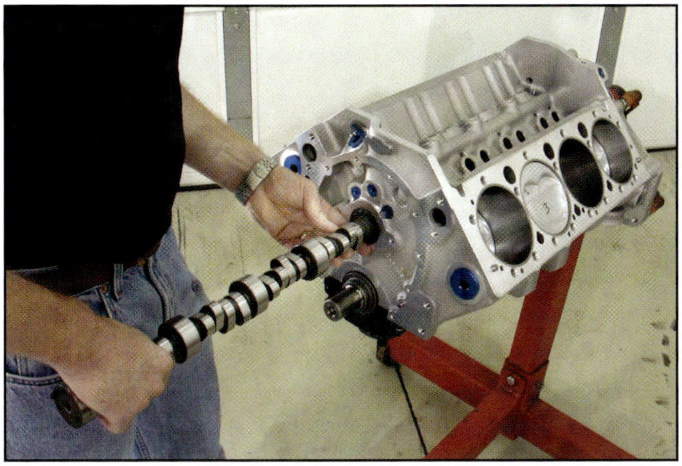

When the Mule was built, there was no other choice than a solid roller cam for this type of engine because of the large cross-sectional runner--racing-bred intake ports get in the way of the standard pushrod location. To get the pushrods pointing straight up again, the rocker arm is offset .450 inch and the pushrod hole is offset another .180 inch in the lifter. These offsets increase the side thrust, which hydraulic lifters weren't really designed to handle…though you can get offset hydraulic lifters today…

The Crane cam is degreed "straight up," neither advanced nor retarded, and is held in place by a Crane billet roller timing chain. The cam sits in the bearings with a tight .0015-inch clearance (cold) to account for the aluminum block's expanding at twice the rate of an iron block. Compare this to the .0025-inch clearance Urban would run on a cast-iron engine to end up with the same clearance at operating temps. Because of this tolerance, I'll have to take a little time warming up the engine (to go to work!) on those cold Michigan mornings to prevent scuffing a bearing.

The Moroso oil pan has a series of kickouts, trap doors and a threaded dipstick bung (arrow) that make it ideal for performance usage…which is why I am using it.

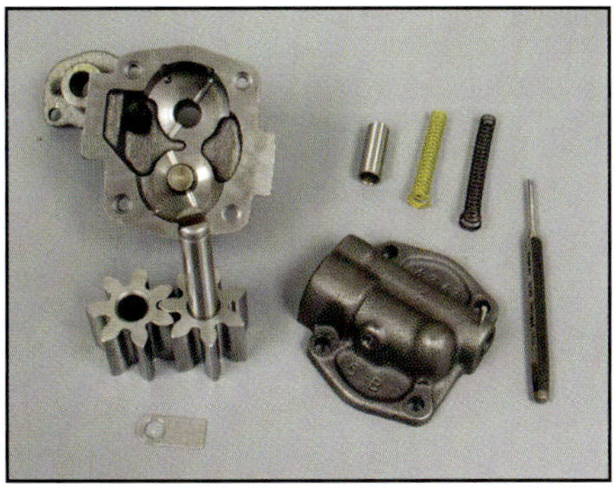

Urban disassembled the Melling oil pump, cleaned it fully and installed a high pressure blowoff spring.

The aluminum block requires 1/4-inch-longer threads on the head studs (3/4 vs. 1/2 inch) than the cast-iron block studs to maintain the required torque rating. This is because the aluminum threads don't have the same load capability as the iron block threads. Notice all the different lengths and thread depths on the head studs alone! ARP is great in these situations as they can tell you what to use based on their experience. The installation method Urban is using (two box end wrenches clinching the blank portion of the stud) is neat, so I captured it for you.

I know the small-diameter, stemmed Ferrea titanium valves are a little overkill, but I like quick revving engines--and lightweight gear helps the engine "yacha, yacha" when you snap the throttle. The head gasket is a .041-inch Fel-Pro composite with a combustion fire ring (everyone calls them the "blue" gasket because they're blue, imagine that...).

Since this head was designed with big ports that have as few turns as possible, many components end up on top of each other--like the studs. The only way to access these is to trim the valvespring seat buckets with a grinder--far away from the engine, of course! (NOTE: Every time Urban grinds or machines something, it is thoroughly cleaned before being brought back into the engine assembly area.) This is not your everyday small-block valvetrain, so a little massaging was required. Once mocked up on the engine, Urban discovered he needed to open up the passage with a grinder where the Crane 3/8-inch pushrods pass through the Chapman heads. Also, the Crower rocker stands and Crane spring buckets needed some grinding to allow room to tighten various head fasteners. This is a high-end valvetrain, but one that will easily handle 7,000-rpm run-ups and flow a lot of air/fuel, which is what I wanted.

The Crower roller lifters and pushrods were large enough and relocated enough that the pushrod holes to be opened up for clearance.

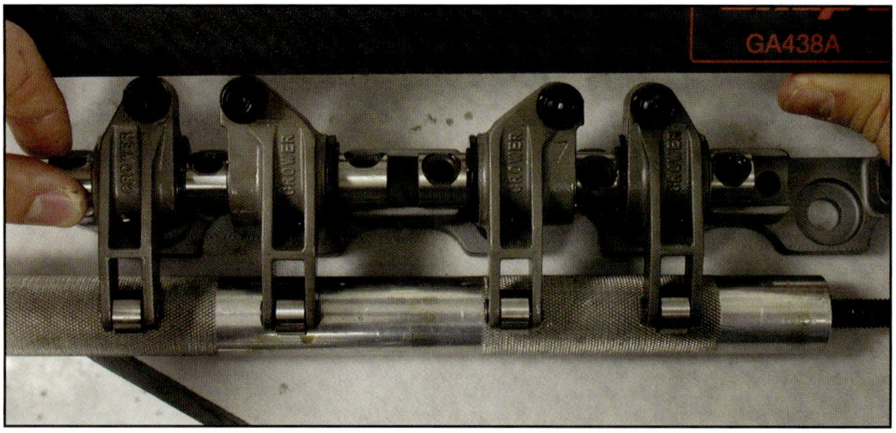

Here are the 1.55:1 and 1.70:1 Crower rocker ratios next to each other. The straightedge really shows the difference between the two sets. I ended up running the 1.55:1 rockers.

The MSD distributor has a sensor on it that provides the EFI unit with a crank position signal. This allowed me to avoid having to install a crank sensor assembly on the front of the engine.

While it is a really boring part, I always run a gear reduction starter as they are small, light and really durable. This one is from Scoggin-Dickey Parts Center.

To mount up to the intake to the 15-degree heads, I created a 3/16-inch-thick aluminum spacer rail, drilled and tapped some 1/4-20 holes, and mounted the spacer pieces with countersunk Allen screws liberally dabbed with red Loctite®.

To control the oil in the top end of the engine, I'm using Billet Fabrication "oiler" valve covers. The covers are plumbed into the oiling system with these Earls -4 braided steel lines to spray a light mist of oil on the valvesprings. This increases valvespring life while reducing the amount of oil in the top end of the engine

I created a 'T' fitting at the oil pressure sender fitting to feed the two valvecovers with fresh pressurized oil. Here, I am determining which AN fittings to use to connect the pressure sender oil to the valvecover fitting.

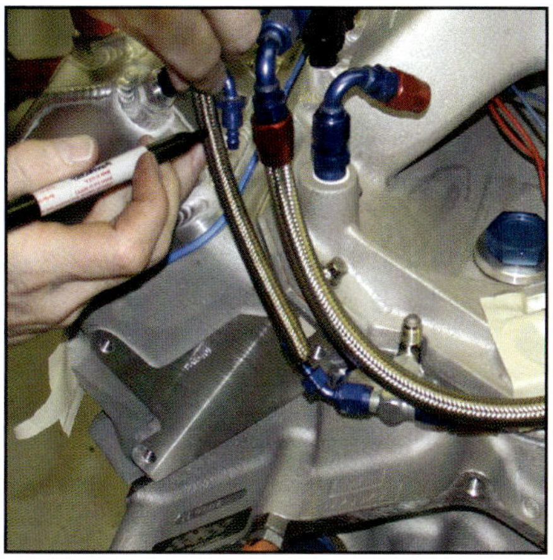

With the fittings determined, I mock up a piece of -4 AN braided steel line and mark it with a Sharpie pen. Then, I cut the hose and assembled the fittings on it before final installing everything with an AN fitting wrench.

I am pointing to the oil pressure feed lines I have created (left hand) and the fuel distribution block I drilled and threaded to allow a separate -6 fuel line to be run to the 100 hp nitrous system (right hand) which never went on the Camaro.

External Oiling Upgrades
As with most aspects of a project, the building of a dipstick tube for the Mule small block seemed like a simple idea. Ha! This is the epitome of 'don't start small projects, because they all end up big!' But it works great and never leaked, which is what I wanted.

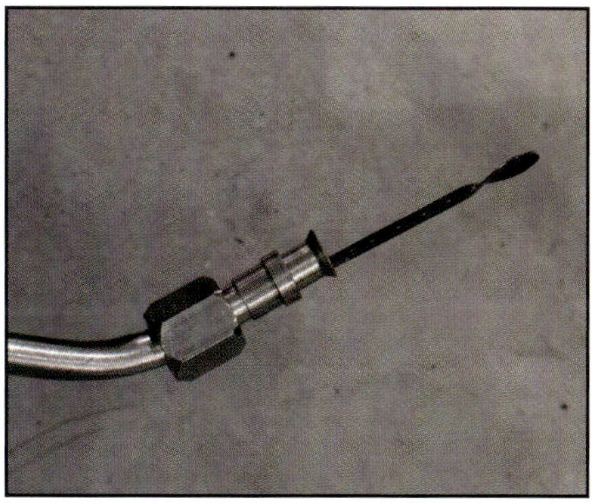

The dipstick tube was bent up out of 0.0.375-inch diameter, 0.065-inch-thickness stainless tube and cut to length based on the length of an existing oil dipstick tube I pulled out of the junkyard. The end that bolts to the oil pan was flared with a ferrel/fitting nut on it to provide a positive sealing point at the oilpan.

The handle end of the Mule dipstick tube was fitted with a 1 inch long piece of stainless tubing that was machined from a solid piece of stainless and press-fit on the tube. This allowed a junkyard find factory dipstick handle to nest in it firmly.

I used a factory GM, V-6 engine, remote oil filter mount to relocate the oil filter to the front of the engine, and put a 'joggle' bend in the dipstick tube to get it out from under the engine.

Here is the other end of the remote oil cooler I used on the Mule. No, its not the most sexy thing in the world, but it worked perfectly through all my abuse.

The o-rings on the fuel injectors were lubed up before being installed in the Flow Technologies—tuned up intake manifold.

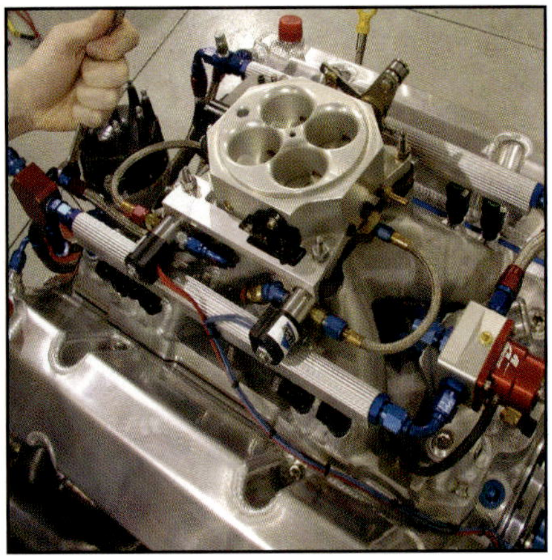

The intake had electronic fuel injector bungs (in each port) and rail stands welded on it by Force Fuel Injection along with many welded-on bungs to mount sensors and fluid lines. I built the nitrous solenoid bracket from 2-inch angle aluminum.

The fuel rails are both fed from the rear and have a return at the front (shown)--I think this maintains more consistent fuel flow to all the injectors versus the full loop (feed in one rail and loop around to return from other rail)

The Holley throttle body being used here doubles as a nitrous plate, which was something I tried with the Mule. It worked well for an initial play. I did upgrade the engine to Kinsler Fuel Injection and then created a 1000 hp twin turbo package to take the powertrain to truly ridiculous levels…

Remember the headers I built to fit tightly on the Mule? Well, here they are on the final assembled engine before it is rolled into the dyno cell. Notice how the headers are level with the bottom of the oil pan—this maximizes clearance and looks the best on the vehicle.

The Mule exhaust was bolted up to the engine on the dyno—but we ran it backwards to clear the dyno absorber. This way, the calibration that was created for the fuel injection was for the entire powertrain that would go in the Mule.

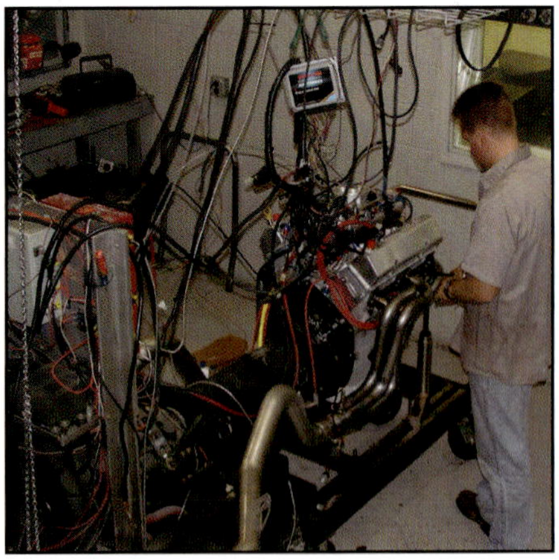

EFI calibration guru Dave Henninger did the initial break-in and cal tuning on the dyno. This is the ideal place to do the start up as any issues (mechanical, electrical or software) can be isolated and resolved without having to struggle through the vehicle architecture.

EFI calibrator Dave Henninger's basic attack is to first drop a "best-guess" calibration (from literally hundreds he has created) into the EFI controller. Then tune the engine based upon the O2 sensor reading of the air/fuel ratio, ignition timing, and manifold pressure at various rpms for maximum power and efficiency. The refinement of the calibration on the dyno is mainly for power production. Once the engine is in the vehicle, the "cal" is tuned for optimum driveability (startup/warmup, idle, cruise, accel/decel, etc.) with two people at the controls (one driving the car and another running the laptop computer). I have used this type of EFI on many cars I've built and know I couldn't go back to a carbureted vehicle--they are so nice once you get them dialed in. Carbs do make great power, however, and do so at the right price.

How would you like to reduce the noise from your performance exhaust by 6 dB for only 12 hp? I'd say them's good numbers! We ran the dyno with the entire Mule exhaust system, which was built by Stainless Works and found their 'quiet' exhaust was almost as capable their 'open' exhaust.

So here it is, the 540 hp small block V8 that first powered the Mule. It was a great engine for this ground breaking vehicle and I would recommend it for anyone.

I'd like to thank Dave Henninger (left) and Kurt Urban (shown with me on the right) for all their help in getting this engine put together and running great. I believe finding great engine builders and tuners is key to ending up with a vehicle that is both fun and impressive—and these guys were pivotal in the Mule being both of those!

Kinsler Injection Upgrade

Making a mid-stream change that needed to happen

OK, some rules aren't as hard and fast as they may seem: I told you to decide on a package and then build the car that way. But now I'm going to show you a major mid-stream change I made on the Mule…and why it was an excellent decision for the car. This would fall under the category of 'Things Change', or, to add more detail: there are times in our lives when you just have to make a change from what you thought was going to work (either visually or mechanically) because it is painfully obvious that the first idea is not going to work. That is what happened here…

Almost as soon as I got the Mule's 400 ci Gen I small block V8 engine running on the dyno, I decided to chase a dream I always had: top it off with an individual runner Kinsler Fuel Injection system. To me, the look of the individual runners is as sexy as it gets—but I wasn't willing to deal with the driveability issues these mechanical fuel injection, 'race-only' systems had always exhibited…but that changed with the ability to set these systems up with electronic fuel injection. Once Kinsler was offering one of these systems with EFI, I knew it was now or never!

This is the picture of performance. The Mule V8 made a huge visual leap forward with the Kinsler system installed. Getting to this point took some work, but it was all worth it!

The System

Individual runner (IR) fuel injection systems have been used mostly in racing and a little on the street since the '30s in the form of mechanical fuel injection systems. The most simple way to describe mechanical fuel injection is as a controlled leak. At WOT, mechanical fuel injection is excellent at flowing a lot of fuel (especially methanol or ethanol: which requires the delivery of about twice as much fuel to the engine vs gasoline powered engines), but at idle or part throttle, they aren't so good (to be kind).

The reason I felt now was the time to move into the Kinsler IR system was its newfound ability to be upfitted with all the necessary items to convert the system to an electronic fuel injection (EFI) system. Now the system would exhibit the best of both worlds: it would have the look of the wild, racing-inspired intake system, but with street manners. That is not to say installing and integrating this system into the Mule was easy, because it was not. My work focused on getting the system to provide reliable, predictable performance and also keep it all looking traditionally impressive.

The looks were the easiest part of this upgrade. Simply put, lifting the hood on the Mule with the Kinsler individual runner intake and injectors on top of the V8 elicited gasps and wows! This 'look' and the ultimate performance were why I went with this upgrade.

To me this is an important upgrade to show because it details how I kept the visual look of the individual runner injection system as clean as possible, while adding all the required sensors and capability needed to run EFI. Whether you are doing this to an intake system or another area on your Pro Touring vehicle, the goal is the same: honor the look of the older product or system, yet make it 'work' like a modern part. This is how I accomplished that task.

The LS Revolution

As I mentioned in the previous section on the buildup of the Mule's Gen I V8, when the Mule was built in 2001, an all-aluminum Gen I V8 was a common choice for forward-reaching hot rods.

But in 2010, when this book is being written, there is no other choice than to run an 'LS-based' Gen III or IV V8 engine from GM.

Part of my change to the Kinsler individual runner intake (shown here) was driven by a sense I had that the single throttle body intake I originally chose for the Mule V8 was looking a little dated. The Kinsler intake was my dream intake for years and it added the required sex appeal I needed on top of a powerful, but somewhat staid-looking, Gen I small block V8.

I was immensely happy with my decision—and even moreso now looking back…as it was one of those things I always wanted to do and, since then, I have run mostly LS-based V8s in my cars. So it really was my swan song for Gen I V8-power.

If I was running a Gen I V8-powered car today, I would have a Kinsler intake topping it off as those stacks just stop traffic when the hood goes up…and I'm glad I built a car with one!

At the same time of the Kinsler EFI upgrade, I installed Summit Trick Flow 18° aluminum cylinder heads on the Mule engine instead of making a ton of changes to the Kinsler system to fit the 15° Chapman heads. This required modifications to the heads for valvetrain clearance fitment, but was a relatively straightforward change since the pistons I originally ordered were for 18° heads.

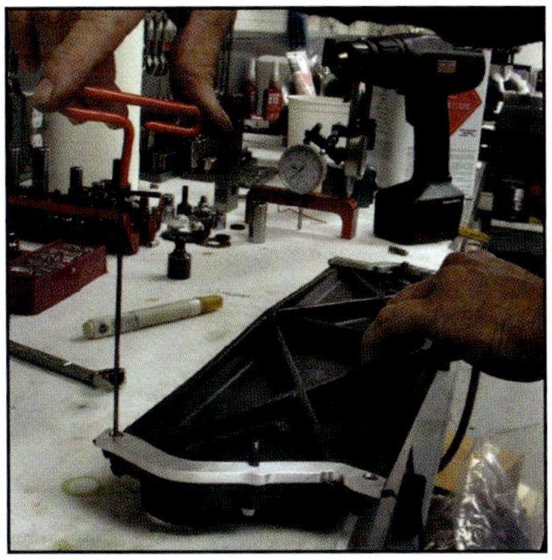

A set of 0.250 inch thick aluminum spacers were machined up to match the shape of the Kinsler valley plate using the gaskets as a guide.

To hold the spacers 'in position' on the intake manifold, two holes were drilled into each spacer and smaller holes drilled and tapped into the Kinsler valley plate so they could be held in place with countersunk ¼ - 20 Allen bolts. Another bolt was threaded from the top down through the spacers to further located the spacers on the valley plate.

If you request it, Kinsler will equip their system with a 'throttle position sensor' or TPS (arrow). This sensor can be plugged into the EFI wiring harness (plug in hand) and send a signal to the engine controller about where the throttle blade positioning is at all times. This makes tuning the car much easier!

Each runner tube was drilled and tapped by Kinsler so I could link up all the runners into a common vacuum container (top, small fittings) and also so a common Idle Air Control (IAC) solenoid could be used—the big fittings with the red caps will be connected to the IAC. The IAC bleeds additional air into the engine (mostly during startup and deceleration) to keep the engine running.

The throttle linkage on individual runner intakes is almost always custom: and the Mule linkage is no different. Kinsler sells the pieces shown here so you can build a linkage system for your application. The parts aren't cheap, but they go together smoothly and help to make sure you have a system that is safe...a critical concern when cracking open eight blades at a time!

The main improvement for this Kinsler system is the addition of the 36lb/hr electronic fuel injectors and extruded aluminum fuel rails (arrows) to the outer face of the injector stacks. They accomplish the job of converting this system to EFI without taking away from the vintage look of the injection system.

This shot gives you an idea how little room there is between the runners and how I plan on positioning the vacuum block on top of the fuel distribution block in the center of the engine intake valley. Also notice that the coolant outlets are in the form of two 120- and 70-degree -10 AN fittings: these will be plumbed to a common thermostat housing on the side of the engine.

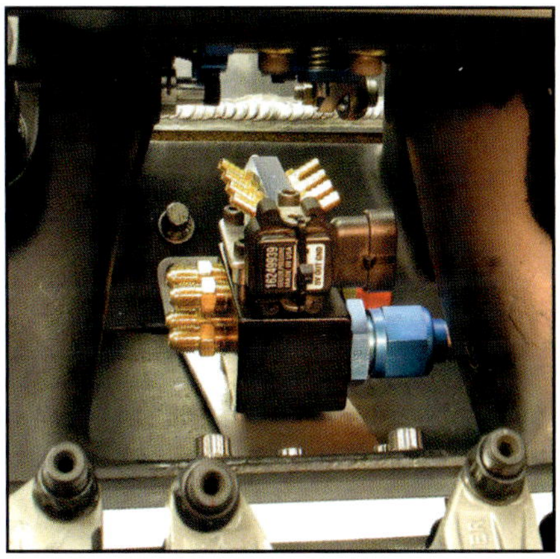

This is a closeup of the custom made vacuum block that all the runners will be connected to and the Manifold Air Pressure (MAP) sensor fitted to the side of it. The aluminum fuel distribution block the vacuum block is bolted to has a -8 AN fuel line running into it and eight -4 AN fittings through which the fuel will run to each individual runner.

The blue block with brass fittings is the idle air control motor manifold that I built. It sits on top of a Kinsler fuel block. The vented sensor in the foreground is the Idle Air Control (IAC) motor.

The completed system ended up with the area between the two runners very crowded with all the hoses, sensors and linkage I mounted in there. Because of this lack of room, I had to try a few different positions for the hoses before I got everything to fit and work properly. This is a common situation: you don't know what you don't know until you try it, then you find out what doesn't work and what does. Seems simple to say it, but it is a frustrating way to find the solution (though it does work)!

Since the coolant outlet on the Kinsler system was in the form of two AN hoses, I built this small collection tank on the side of the engine to mount the thermostat and housing to, along with the two coolant temperature senders (one for the fans, the other for the computer/gauges). Notice how I have cut off the outlet for the upper radiator hose, 'reclocked' it, then welded it back on so it's pointing at the radiator inlet fitting.

Here is how the front of the engine ended up looking: this was one of the first Vintage Air Front Runner accessory drives combined with my coolant assembly for the Kinsler system.

After building something so it looks great, you need to spend the time to make it 'work'. In this case, that process starts with getting each throttle to start out in the same position in the intake bore. This is accomplished with a small 0.001 inch feeler gauge and a lot of time adjusting the shaft angle with the linkage.

The initial testing and refinement was performed on an engine dyno with skilled engine builder Kurt Urban manning the controls and tuning the hardware.

The dyno is a great place to dispel rumors and understand performance gains/losses from mechanical changes. To that end, once the engine combination was getting locked down, we tested some different length stacks to see what it would do to power output and rpm band placement: the engine liked the longer stacks…

Ahhhh, the sweet sound of a small block V8 at WOT pulling like a ton of bricks! Nothing like it in the world…

The final package that was installed in the car consisted of the Kinsler individual runner intake with the mid-height stacks nestled inside K&N air filters. The stack and air filter height was really set by the hood clearance of the Camaro—as I would have liked to run taller stacks, but they wouldn't fit.

Here is a shot with the air cleaner lids removed to show the stack location inside the air cleaners. Also, notice the inlet air temperature sensor (arrow) nested in one of the air cleaners—this provides inlet air temp information to the engine controller so it can chose the appropriate air/fuel ratio and ignition timing for the current inlet temp.

Swapping Heads

Obviously, swapping heads before the engine has ever been bolted in the car is tough to justify: what can I say, sometimes you just have to jump in with both feet. In this case, it was really tough to pull the Holley throttle body system and 15° Chapman heads off the engine, but it was the only way to get the Kinsler system bolted on without some really serious rework to the injector system and the engine.

Let me explain…The Kinsler system was built for 18° heads, so swapping on a set of 18° Trick Flow heads made the Kinsler system a bolt-on. While the Trick Flow heads required some rework to fit the valvetrain, engine builder Kurt Urban was well versed in those changes and made them with little fanfare.

This is not uncommon theory when building a Pro Touring car: you are constantly weighing the cost/complexity/capability that will result from making a change. In this case, I determined swapping heads was the least expensive/least complex solution, while maintaining the power production capability of the engine.

Making the change to 18° cylinder heads wasn't that hard, but trying to mate up the 1.55:1 Crower rockers with the big 3/8 inch pushrods was a handful! See arrow for mismatch of locations. Luckily, engine builder Kurt Urban wasn't fazed and we went through the needed steps to get the Crower rockers and pushrods working together.

The first step was to mill out clearance in the pushrod holes on the Trick Flow 18° cylinder heads. This was accomplished by mounting the heads on a mill and running a ½ inch diameter bit down into the head material. This is a tricky step as taking too much out of this area can turn the heads into junk—this is where having a skilled and experienced engine builder pays dividends.

The next step was to insure there was sufficient valvespring height to handle the valve lift. Step one of this check was to install the valve seals. These are lubed up, started by hand on the valvestem and then pushed onto the stem with an install tool.

Urban hand checked every spring height by installing the spring bucket shim (which has been clearanced around the head bolt), stem seal, retainer and locks on the valvestem and measuring the distance between the bucket shim and retainer. This dimension was then written on the head with a black Sharpie pen (see arrows). The heights were made equal via shims under the bucket shims once all the dimensions were known.

A simple check of the valvespring heights once all the shimming is completed is to place a straightedge on top of the valvespring retainers across the length of the head. The straightedge should be perfectly flat across the retainers. If it is, the valvespring height work is complete.

With the heads final installed (torqued, valvetrain lash set, etc), the valvecovers and gaskets needed to be clearanced to fit around the pushrods. This was accomplished by some judicious grinding on the valvecover mating flange and Fel-Pro gaskets—take too much and you create a 'leaker', too little and stuff makes bad noises: so this requires some time to get it all 'just right'.

Turbo Power

This is where the Mule powertrain went haywire... 1000+ hp twin turbo!

Well, this wasn't really done when I owned the car, but I want to show the Mule being upgraded to twin-turbo power because it is how the vehicle exists today. When I finished the Mule, it was a nice car and my friend Charley Lillard decided he had to have it...so I sold it to him. Once Charley owned the Mule, he felt it needed more power—and he knew I had an idea of how to install a twin-turbocharged engine package that might possibly make 1000+ hp at the flywheel. So, this section shows how we upgraded the Mule to ridiculous power (and how you can do it, too).

To me, this 1000+ hp twin-turbo powertrain was the final step needed to 'make' the Mule—driving this car in 'anger' now is a true thrill and I am always impressed at how well the Mule handles the power.

New pistons, crank, rods, etc...

The surprising thing about building a big hp turbo engine is that it's not like building a supercharged, nitrous or big cubic inch engine. Turbos, when properly controlled and fueled/sparked, are quite 'easy' on parts. But, there can be an immense amount of cylinder pressure applied to the piston/head gasket, etc...so the piston needs to be able to handle that pressure and the resulting additional stress of preignition/detonation, when they occur (notice I didn't say 'if'...when you are building a twin-turbo engine package, it is a given you will be applying boost to the engine!).

To handle this pressure, engine builder Kurt Urban made the call to switch to a piston design that had more material between the top of the piston and the first ring land and also had the piston pin below the oil ring. To do this, a new shorter throw crankshaft was needed, as the prior piston design was for a naturally aspirated (NA) engine: It had a narrow slice of material between piston top and first ring land and the pin running through the oil ring pack—which isn't good for a turbo motor. This also drove the need for a new set of rods, different camshaft, Inconel exhaust valves (to handle the increased exhaust temperatures) and larger fuel injectors to flow more fuel. A 0.055 inch thick multi-layer steel head gaskets were also added to the engine package.

The block and heads remained the same, but the top half of the Kinsler individual fuel injection system was removed as two common plenums with custom throttle bodies on each one were built to direct the boosted air into the engine

As before, the engine is buried in ARP fasteners to hold it all together (even more important now with the power output) and a bunch of work went into protecting for the immense underhood heat that results from running a turbo system. This included building heat shields and venting.

Top 5 Custom Turbo Header Building 'Truths'

1. *There is no 'one correct way' to do a header, but there are a lot of 'wrong' ways! Be prepared to cut off and start over on at least one section of the header(s)*

2. *Use quality bends, flanges and merge collectors from a company like SPD or a similar company so the manifold is durable, flows the maximum exhaust and looks great*

3. *The welds should be TIG (tungsten inert gas), the bead tight, the penetration thorough and patience the order of the day to avoid warping the entire assembly into a pretzel*

4. In general, what looks 'smooth' usually is what flows good—rough transitions, sharp angles and hard lips will impede flow and hurt power output

5. This all takes time: time to build headers, time to learn how to build headers and time to learn what not to do when building headers. If you don't have the time, pay someone who has put in their time.

This is the Mule's turbo package just after it came off the engine dyno where it made 1021 hp at 7000 rpm and 810 lb-ft of torque at 6500 rpm at 16.3 psi. Take a look at the dyno sheet: the torque is over 700 lb-ft at 3800 rpm and never comes down!

The intercooler for the Mule was a challenge to make big enough to properly cool the intake charge and yet package inside the stock '69 Camaro envelope. To make it work, the factory radiator core support was heavily modified...this section shows how that work was completed.

Power Output

Here is the best run dyno output for the twin-turbocharged, intercooled, 380 ci Gen I V8 small block that powers the Mule today:

RPM	Trq	Hp	Boost (psi)
3000	526	301	7.1
3100	543	320	8
3200	563	343	8.8
3300	589	370	9.9
3400	617	399	11
3500	643	428	11.8
3600	666	456	12.6
3700	688	485	13.3
3800	704	509	13.8
3900	710	527	14.2
4000	709	540	14.5
4100	713	557	14.8
4200	712	569	15
4300	719	589	15.2
4400	718	602	15.4
4500	717	614	15.6
4600	722	633	15.8
4700	734	657	16
4800	741	677	16.2
4900	745	695	16.4
5000	754	718	16.4
5100	761	739	16.5
5200	763	755	16.5
5300	768	775	16.4
5400	779	801	16.2
5500	784	821	16
5600	779	830	15.9
5700	779	846	15.9
5800	778	859	15.9
5900	780	876	15.9
6000	787	900	15.9
6100	793	921	16
6200	801	946	16
6300	805	966	16
6400	807	983	16.1
6500	810	1002	16.1
6600	810	1017	16.1
6700	796	1015	16.2
6800	785	1016	16.3
6900	777	1021	16.4
7000	766	1021	16.3

All power numbers are corrected. The water temperature was 150° F, the oil temperature was 162° F and the oil pressure varied between 73 and 82 psi over the course of the sweep test.

By the way, if you ever get the chance to be at a dyno during a sweep test on a thousand horsepower turbo motor, do it: it is amazing to watch and hear the engine thrushing its way to four digits on the dyno!

The 1 ¾ inch primary exhaust headers, turbo mounting structure, 2.5 inch pressurized air tubing and 3 inch exhaust system were all custom fabricated by Denny Dera, Whitey Brown and Rick Rasmussen (who at the time worked at the now-defunct Wheel to Wheel Powertrain).

This turbo system was made to 'work'…so all the tubing was TIG welded at the butt-welds, all temperature sensitive areas are protected with metal heat shields and stainless steel bolt clamps are used on all the tubing mating points.

To maximize durability, the turbos are fed pressurized engine oil to keep them lubricated/cooled [red arrow] and the vacuum controlled wastegates [blue arrow] were positioned as close to the outlet point of the turbo as possible (this minimizes a delay in pressure control).

Looking back at this engine package buildup all these years later, I think it was a fitting requiem for the Gen 1 small block V8. While the engine internals were upgraded to handle the turbo power, it is still a Chevy small block V8 handling 1000+ hp of power on the street—impressive.

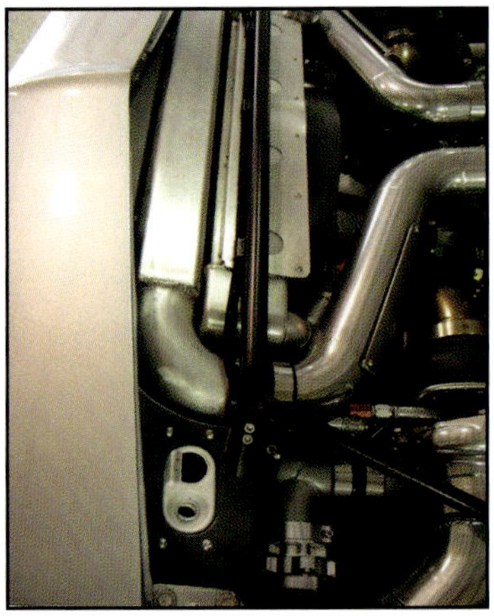

Here are the mods that were performed on the radiator core support: the radiator/intercooler/condenser assembly was pushed forward 2 inches, the center portion of the core support was cut out and a new structure fabricated. Also, the stock hood latch was replaced with two latches on the ends of hood.

The Mule engine bay is truly impressive when the hood goes up on the twin-turbo V8. It runs smoothly and the exhaust note is very quiet compared to any other car I have owned (I like turbo whoosh, but it is very different for me…).

I am big on having a visual guide when building something, so for this engine package, I used this photo of a big block 'mystery V8' road racing engine that was campaigned in the '80s.

Of all the engine packages my cars have had, this one is in the Top 3. Doesn't it look impressive? Everything on it looks powerful and yet, it looks somewhat familiar. This is a street-ready, 1000+ hp '69 Camaro that will still gladly run all day at a racetrack! It still amazes me…

This is the twin-turbo Mule V8 as it is going together. Ace fabricator Whitey Brown is in the process of fitting/cutting/welding all the tubing into position (by far the most expensive, complex aspect of building a turbo engine package).

The best way to build turbo tubing is to do the fitting/cutting/tack welding in the car, then move the tubing assembly to a mockup engine on a bench (like this one for the Mule) so you can easily weld all the joints together.

I had originally planned on mounting a downflow intercooler (right) in front of the Griffin radiator (equipped with trans and power steering coolers in the side tanks), but it just wouldn't fit correctly under the hood and look right—so we changed plans.

When starting on a turbo engine package, the most important step is selecting where the turbo(s) are going to be mounted. For the Mule, I wanted the Precision turbos up front and on top of the engine bay—as I think it looks most impressive.

To hold the SPD exhaust header collector in position, fabricator Denny Dera cut/bent/drilled up a simple piece of 1 inch x 0.095 inch mild steel strap, then bolted it all in position. This way, the collector location wouldn't move around as the tubing was fabricated to fit.

A SPD exhaust flange plate was then bolted to the 18-degree Summit cylinder head with 12-point ARP exhaust bolts. The next step after that was to determine where each primary exhaust tube would run into the collector (note the #s written on the collector).

This custom-built exhaust header is the result of about 10+ hours of fitting/cutting/tack welding by an experienced fabricator using SPD 1.875 inch diameter T304 stainless steel primary tubing bends.

The tack-welded together header was rough fabricated in the Mule, then moved to the bench-mounted engine with the collector resting on an upright in prep for final-welding the mating points.

The SPD T321 stainless steel collectors have 1.875 inch diameter and 15 inch long primary tubes—and use SPD's double slip collector add-ons to minimize exhaust leakage at this critical joint.

The SPD double slips capture the primary tubes on both their inner and outer faces, which seals the mating joint as solidly as possible while allowing sufficient 'slip' to account for the excessive expansion/contraction experienced with turbo headers (they will glow red on occasion).

This is the SPD collector tubing/double slip/primary tube mating point (finger) in position on the turbo exhaust header.

Once the primary tubes were completely welded together (using patience to avoid putting too much heat in the tubing, warping them), the header was clamped to the bench and the tube/flange mating point final TIG welded.

The 3 inch diameter dump tubes that run out of each turbo were routed through the bottom of the engine bay and mated to the existing exhaust system. SPD sealing flanges were welded to the ends of the tubing so 'V-band', or 'Marmon', clamps (arrow) could be used to minimize exhaust leakage.

At this point in the build, it is becoming apparent space is going to be at a premium to package the radiator/intercooler/condenser in the front of the car (note the stock rad support is still in the car).

From the moment I first saw them to today, I really like the air inlet boxes built by Whitey Brown for the top of the Kinsler fuel injectors. They look purposeful, fit under the hood and, above all, work perfectly.

Here we are trying to package the radiator in front of the turbos. While I could have let the radiator fitment drive the location of the turbos, that didn't seem to make sense. So I knew we'd have some fitment issues, but not as much as we experienced…

The wastegates (arrow) were positioned off the front of the turbos, so we thought they were what was preventing the radiator from fitting in the car (ahh, if only life is that simple).

After considerable thought and playing around to see what we could move around on the turbos, it was decided the radiator core support would be cut apart and a cross-flow intercooler would be used. This allowed the turbo assembly to remain as shown.

Fabbie Whitey Brown then welded together an 'upflow' intercooler by adding handbuilt aluminum tanks and tubes to a 22 by 14 inch Spearco air-to-air intercooler core. The machined aluminum inlet tubes were 2 inches and the outlet tubes 3.9 inches in diameter.

As a tip, always cover finned areas on coolers with thick sheet cardboard when doing fab-work on them—this minimizes the damage to them during this clumsy work. And notice that we are trying to use the stock rad support here…

The original idea was to open up the stock rad support hole and slide the radiator/intercooler/condenser assembly forward through it…but we kept cutting away pieces to fit components and tubing until it really didn't exist anymore—so we fabbed up a new crosspiece.

As often happens, you keep cutting until there is nothing left. Here, we have cut away the center of the rad support and built 1 inch round tubing cross braces to support the radiator (arrow). Whitey Brown is welding mounts on for a more compact rack & pinion steering system I have upgraded to…

The intercooler and A/C condenser were mounted using ½ inch diameter, 0.060 inch wall thickness round mild steel tubing (arow) bent to mount to the remnants of the stock radiator core support.

Here is a closeup of the heat exchanger 'stack': the condenser is up front (on right), the intercooler in the middle and the engine radiator at the back (on left). This is why I went to a bigger Griffin radiator—the air that gets to the radiator will be quite hot after shedding heat from the other two exchangers!

The top of the radiator support ended up with a bent 1 inch x 0.090 inch wall mild steel tube that mated with the stock core support via two short spuds held together with two Allen bolts/side. Also, note the bent aluminum tubing welded to the outlet of the intercooler: nice!

This gives you an idea how every inch of space gets used! A custom fabricated -16 AN radiator inlet tube has been welded on [red arrow] and the pop-off (intake side) and wastegate (exhaust side) [yellow arrow] are just below the -10 AN braided line for the oil cooler inlet [blue arrow].

The air inlet tubes for each turbo run into the fenderwell area and ended up having closeout plates on the fenders to draw cold air from the front of the vehicle.

The passenger side exhaust manifold turbo mount (available from SPD) tips up slightly and is fully supported by a ½ inch round tubing structure. This positions the turbo in a similar clocking to the driver's side turbo and the structure keeps everything from sagging due to the extreme heat.

This is the passenger side turbo in position on its support and exhaust flange.

Here is Whitey Brown test fitting one of the intake plenum's he welded up from sheet aluminum. Notice how we are using just the lower portion of the Kinsler individual runner intakes to achieve a lower intake height—possible since each plenum has its own throttle body.

The throttle bodies were machined from 6061 aluminum and use Kinsler throttle plates and hardware—these are built to control the 16+ psi boost pressure coming at them!

Here is the twin-turbo engine bay taking shape in the Mule. I like the symmetry of the tubing and components and how everything seems to 'make sense' in its placement—and that it all made 4 digits of power (over 1000 hp)!

The biggest problem with a turbo system is the incredible heat that comes off the exhaust tubing. For this reason, heat shields will rest on the subframe downbars, a heat resistance coating was applied to the valvecovers and the hood will be vented to allow air to pass through.

I like this shot because it shows how the business of building a custom exhaust works: it's a 'Cut and Paste' business. This is the first version we did of the passenger side manifold. Look at the final exhaust and you'll see we changed the turbo mounting to get both turbos to look 'right' in the engine bay—one of many changes!

This is where having a team that understands the heat issues with turbos comes in handy: the brake booster/master cylinder have adequate space between them and the exhaust tubing.

So here is the turbo system almost fully installed in the Mule: notice how far the intercooler sits in the front fascia and the metal heat shields on the front frame section downbars. The system will now be fully removed so everything can be polished/plated/ painted.

Pulling the entire twin turbo system out of the Mule is a real feat: there are a lot of parts! The tubing was coated with a heat resistant material, the turbo housings sent out to be polished and ARP fasteners finalized.

Then the engine was pulled out and sent to the dyno to be broken in and a base calibration put into the electronic controller to be able to start the engine, have it idle and run under part throttle.

Here is well-known LS V8 engine builder Billy Briggs checking the turbo Mule engine in prep for running it as a NA (normally aspirated) engine. The 'good' turbo guys run their engines NA first to get them running right. A rough calibration was established to start and idle the engine, then the turbos added.

With the engine testing (see Dyno Sheet for power output) and the appearance work complete, everything was reinstalled in the Mule and driving development initiated to complete the calibration work. Looks good, doesn't it?!

The fiberglass Harwood hood that had been so beautifully prepped was pulled for a steel hood - as I needed to add vents in it to allow more air to flow through the engine bay. We chose to position the vents over the top and just behind the turbos to promote the maximum airflow.

To get an idea of some ideas for vents, Brian Baker, from GM Design, came by and laid down some tape on the hood. This is an excellent way to visually discuss installing vents in a hood.

This shows some of the ideas that were thrown out: equal length vents, varying the length of the vents to mimic the shape of the '69 Camaro hood and more…

And this is what we ended up with: 11 equal length vents punched into a steel hood (it was determined the fiberglass hood would never hold up to the intense heat from the turbos).

The vents are punched down to suck the maximum amount of air out of the engine bay.

The shape of the vents in the hood are similar to the production-car vents located at the base of the windshield, so they visually fit in quite well on the car.

Clutch Setup

Making sure the clutch is up to the task of transferring the power

I love to build cars with tons of power…who doesn't? But for me, that power is worthless if it can't be effortlessly dumped to the pavement. For that reason, I have spent a lot of time getting my clutch to transmission package to the point that it just goes in the car and works first time. Here is the setup I did with the Mule.

There are a lot of people that focus on making big power but never think about the details about getting that power to the ground. This section will show some of the details I go through to make my cars go, like using ARP fasteners and a Centerforce clutch.

Twin Turbo Clutch

The clutch shown on these pages was for the normally aspirated version of this engine. A twin disc clutch rated for higher power levels was installed to be able to handle the 1000+ hp of the twin turbo engine.

I like my clutches to have a very even, medium force clutch pressure, but be able to handle extreme power. For that reason, I use the Centerforce Dual Friction clutch with a Centerforce pressure plate. The additional surface area of the dual disc clutch allows it to hold the engine power, but not require insane spring pressures to do so.

'Bedding in' the clutch

Clutches are like brakes in that they need to be 'bedded in' so they provide consistent, long service. The way I bed in my clutches is to do as few low rpm clutch engagements as possible while driving about 500 miles—usually trying to achieve these miles in long stints on the highway.

From what I am told, this helps the clutch to transfer some of its material to the pressure plate and is what makes the mating point 'work'.

Whatever is happening, this has provided an excellent starting point for durable service from my clutch combinations, so I continue to do it

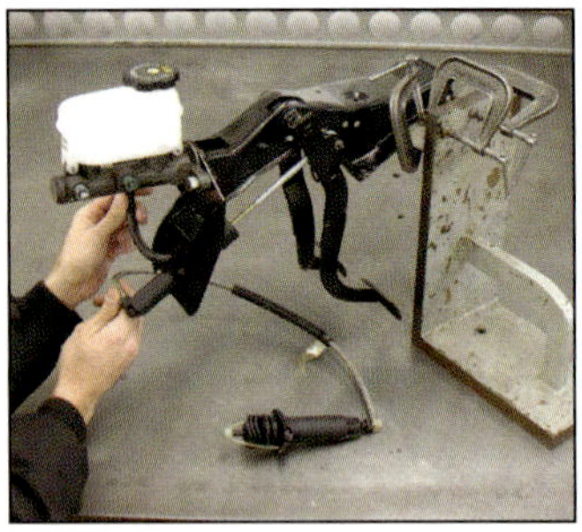

While there are other options today, back in 2001 if I wanted a production-style hydraulic throwout bearing system, the best option was to use a 2001 Camaro assembly (pn 12570277) and build a custom bracket to hold it in position on a 1969 Camaro bracket (and achieve the required 6:1 pedal to motion ratio).

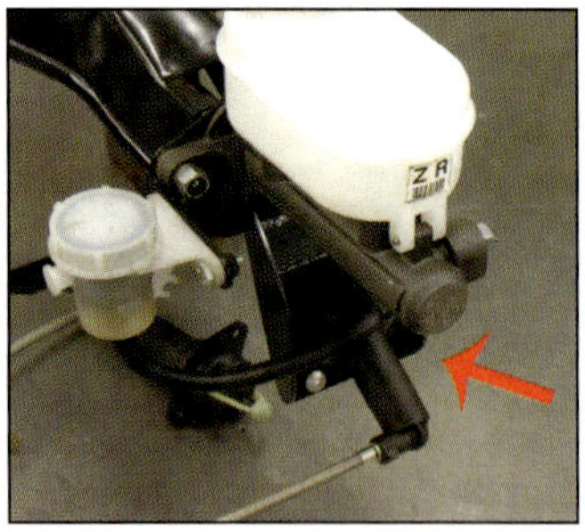

This image shows the adapter bracket I built (arrow) to hold the late model slave cylinder and reservoir in position on the 1969 Camaro brake/clutch pedal bracket. DSE now offers a bracket like this.

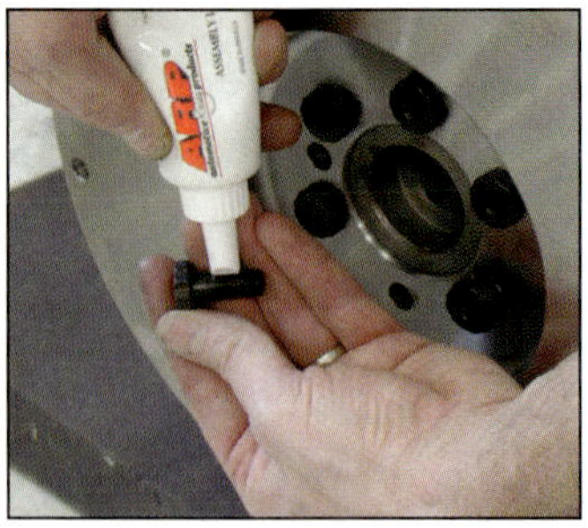

The first step to assembling the clutch and transmission on the completed engine is to wipe clean the crank flange, then install the Center Force flywheel on the flange, apply ARP lube to the threads and hand start the 12-point pressure plate bolts into the crank flange.

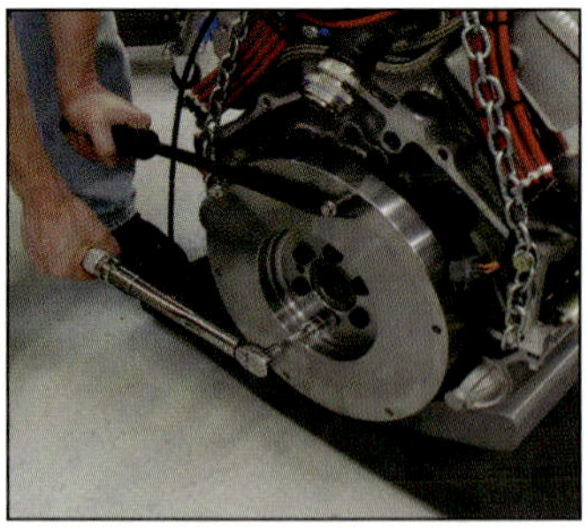

The next step is to torque the ARP fasteners to 70 ft-lbs. I use a large pry bar wedged between two clutch mount bolts to hold the pressure plate from rotating, then apply the torque to the bolts in a 'star' pattern.

I then wipe down the pressure plate with a clean rag soaked with brake clean to prepare for installing the clutch. This needs to be as 'clean' as you can get it—especially of any grease or silicone-based materials.

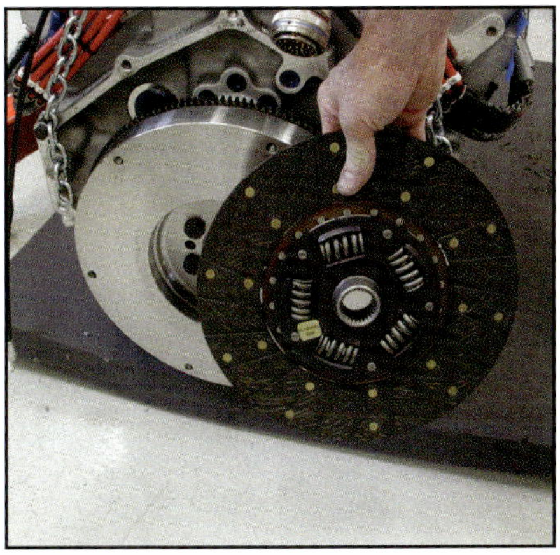

The clutch plate is then positioned on the pressure plate and a pilot shaft used to center the clutch on the surface plate.

The Centerforce clutch is then positioned on the pressure plate. I apply two drops of red Loctite to the threads of each fastener and hand start them into the pressure plate. I then tighten the bolts just past hand tight, which allows the splined pilot shaft to be wiggled out.

I then tighten all of the clutch fasteners a little at a time in a star pattern until they are all at 35 ft-lbs of torque. Do this relatively quickly as the Loctite does start to take a 'set' and can 'false' the torque reading if you wait too long. The clutch is now installed.

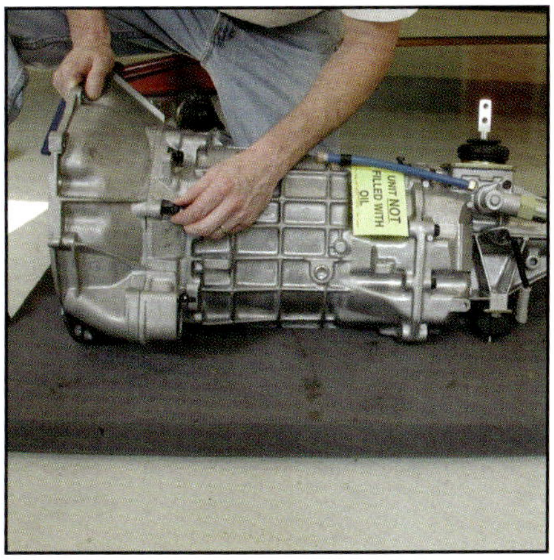

The next step in assembling the engine/transmission combo is to make sure the T56 bellhousing is properly installed. On these late model transmissions, the bellhousing's bolt on, so you need to make sure you have the bellhousing/transmission bolts torqued to 35 ft-lbs all around the bellhousing.

The transmission can be slid on the back of the engine, with the splined input shaft of the transmission feeding into the clutch—apply a light amount of engine oil to this shaft and carefully feed it into the clutch. If you have the splines aligned, it should slide in with a little resistance, but cleanly. Make sure the throwout bearing arm (in hand) is properly seated on the bearing by wiggling it—it should move around slightly (if it is not properly installed, it will be firmly locked).

Again, I like to use the GM production T56 throwout bearing assembly, which comes as a unit that I install on the transmission at this time (as shown). This system works great every time, which is a big help as dealing with throwout bearing issues is brutal: the trans needs to come out every time you want to work on the clutch/throwout bearing assembly. Believe me, this is the easy way to do a clutch/throwout bearing setup!

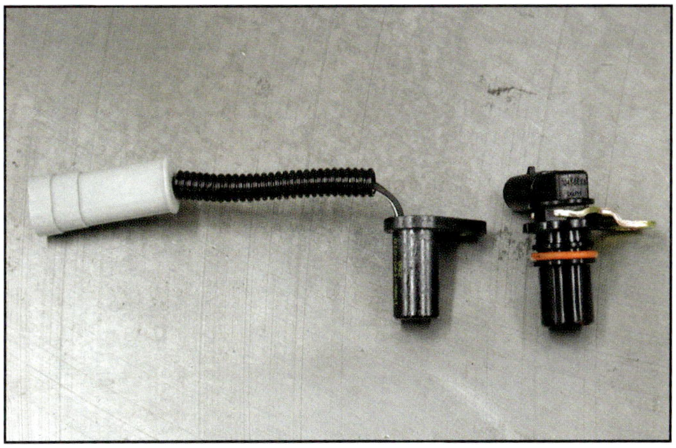

I have my T56 six speed transmissions upgraded with Viper internals by D&D Performance to be able to handle the power and abuse I put to them. To do this you need to replace the factory GM speed sensor that goes in the T56 transmission (on the right) with the Viper unit needed for the Viper main shaft and tail housing (on the left).

Transmission & Driveshaft Prep

Taking steps to get the power to the rearend

Obviously, the engine is the 'sexy' part of the powertrain—and the one that usually gets most of the money and love. But as you all know, that power will mean nothing if it can't get to the ground. To that end, my goal when selecting and integrating the transmission/driveshaft/rearend components on my cars is so they will provide reliable service with at least 30% more power than I am planning on making. That might seem excessive until you consider that with the Mule, we started with a 600 hp engine and ended up with a 1000 hp engine, so I am glad to have built in some added capability!

Driveshaft

I used an aluminum driveshaft from Inland Empire Driveline, in Corona, CA, on the Mule and have never had a problem with it through all these years of abuse (the Mule's owner, Charley Lillard, likes to take the car to 'track days' and occasionally invites me along: we have thrashed the Mule all day long and the only work we have done is to add more gas!). While carbon fiber driveshafts were hitting the market as the Mule was going together…and I know using one would have helped me remove some weight from the vehicle…I decided to go with aluminum because I have had great reliability over the years and the weight wasn't that much better.

Measuring for the driveshaft is pretty straightforward: once I have 'final installed' the transmission and rearend and set the vehicle at 'ride height', I measure between the centerline of the u-joint mounts on the transmission output shaft and the rearend. With the Mule, I provided this distance to the team at Inland Empire Driveline and requested a 3.5 inch diameter aluminum tube for the driveshaft with forged aluminum yoke/u-joints to be able to handle a 1000 hp engine. I also spec'd out that the yoke have 30 splines on it to mate to the Viper-prepped T56 trans.

When A Package Arrives

When a product shows up at my shop, I test fit it as soon as possible (this is a key issue: too many enthusiasts receive a product and set it on a shelf for months…or longer). If you are building a project car, I suggest ordering the parts early enough that they show up at least a few weeks before you need them. But that doesn't mean when they arrive you just set the unopened box on the shelf…you need to pull the part(s) out of the box when they arrive.

This seems like the basics, but I strongly suggest you pull parts out of their shipping boxes, inspect them and then test fit them as soon as possible. This way, if the part was damaged in shipping or if the part was built to the wrong dimensions (we know you measured correctly, right!?), at least you figure it out as quickly as possible and can contact the manufacturer to remedy the situation.

Whatever parts that get shipped to you, I suggest pulling them out of the boxes as soon as they show up so you can inspect them and test fit them.

Transmission

As with many of my cars, the Mule was built with a D&D Performance prepped T56 6-speed manual transmission. These late model transmissions have shown themselves to be fully up to the task of consistently multiplying the power. While I run lightweight Red Line transmission fluid in my racecars, I run GM manual transmission fluid in my Pro Touring cars as I am most concerned about having a reliable package both on the street and the racetrack.

The Inland Empire Driveline driveshaft was built to be able to handle 1000 hp, which is good as the twin turbo Mule engine puts about that much to the tires! I just oiled up the yoke, slid it into the back of the trans and hooked up the u-joint at the other end to the rearend.

Yes, I use GM Synthetic Manual Trans Fluid (pn 12346190) in my transmissions. It has provided reliable lubrication and cooling: and I am a creature of habit! So I keep on using it. I buy it from SDPC (sdpcparts.com). I have also used Red Line Oil with great success.

Painting the Mule

My Tip: Spend the money & time to end up with a car you can be proud of

OK, I admit it...when I was younger, I thought I knew how to paint a car. But after building my third car, I decided I really didn't know and really didn't want to know what I needed to know to paint my own car. My decision stemmed from the fact that the work seemed amazingly difficult and it felt like it took forever to make any progress. I just decided not to do it anymore. Now, that doesn't mean I stopped learning about the paint process, because once I stopped doing the work, I really wanted to know more about every step needed to do an excellent paint job so I could make excellent customer choices.

My search to improve the painting process has led me to some great shops and process tips, but I have to admit to a few miscues, which we'll get into later in this section. If you know anything about me by now from the process breakdowns I have given in this book, it's that I like to try new innovations and ideas. I work hard researching everything to minimize any issues, but even the best of plans go awry sometimes.

With that said, the paint job on the Mule has withstood the test of time. It looked spectacular the day it rolled out of the paint shop and continues to look amazing today. I believe if you used the process outlined in this section on a car built today, it would look just as good (if not better). There are 98 photos in this section detailing the paint process we used on the Mule, so it should be painfully clear how the paint work was done and how to replicate it. With that said, here is how we painted the Mule.

Paint Body Prep

I have had over 10 high-end cars painted in shops and can say with confidence most paint shops are great at smoothing bodies and applying paint, but not at doing heavy duty sheetmetal replacement work. I'm sure there are the exceptions to this rule, but I haven't found them.

For this reason, I have a sheetmetal shop repair the body, then take my car to the paint shop for them to smooth the bodywork and apply the paint. FYI...

Cad-plated Factory Fasteners

Another trick I have learned from doing a ton of buildups is that the factory body and trim fasteners can often be reused. I do this by sending the fasteners that are in good condition (but rusty) out to be chemically stripped and cadmium plated by my local plating shop, Mid-State Plating (midstateplating.com).

When the fasteners come back, they look brand new. In fact, over the years, I have collected literally hundreds of these fasteners and now have a massive cabinet filled with them. I love using a fastener that was designed for a specific location/application and they are inexpensive to find and plate, which is even better.

So, on your next buildup, once you get the fasteners off your car, send them out right away to be stripped and plated. That way, they will be ready when you start to reassemble the car.

The Mule's Paint

The Mule was painted with PPG paint products. Now, I have had painters tell me paint products are all the same, which seems odd: I believe in the premium brands. Having experience with just about every paint brand on the market, I have found there to be a difference between economy and premium paint—and now use only the best. I consider PPG to be one of the best. The painter of the Mule, Mike Verellen, has used PPG products in his shop for years and swears by them. After I saw the results of his team's work on the Mule, I can tell you I agree with him.

The Mule was the nicest car I had built in my life when it was completed (and it still rates in the Top 3 of paint jobs for cars I have built). When you consider all the time I had in developing the various systems and parts for the car, it only made sense to give it an amazing paint job.

I hope you take the time and put together the money to have the same type of paint work done on your pride and joy buildup. There really is nothing like having a beautifully painted project car roll out of the shop onto the streets!

PPG Paint Products for the Mule

Here is a list of the paint products that were used on the Mule (back in 2001):

Description		**PPG Part Number**	**Mix Ratio** (paint:reducer:catalyst)
Wax & Grease Remover (Acryli-Clean)		DX330	
Reducer (for Primers and 2-stage Paint) (reducers used depending on ambient temp)		DT870 & DT885	
Etching Primer		DP 40LF	3:1:1
Catalyst		DP 402LF	
High-Build Primer		K38	4:1:1
Catalyst		DCX61	
Light-Build Primer		K36	4:1:1
Catalyst		K201	
Primer/Sealer (White)		DP 48LF	4:1:1
Catalyst		DP 402LF	
2-Stage Paint			
Color	(Silbersee Silver)	DBC 37637 H	1:1
Clearcoat		DCU2002	3:1:1
Catalyst		DCX61	

Painting Tips

Fresh Air
Notice the fresh air equipment the painter is using and the fact that very little skin is exposed: this is a requirement with today's paint products as the paint is much more aggressive at 'adhering' to things than in the past.

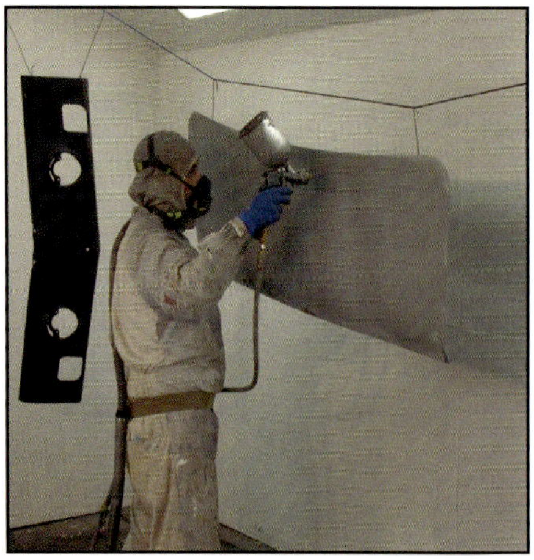

Leave Room
When you are hanging parts, check that there is room to paint by walking around all sides, pretending to paint everything top to bottom. Leave yourself more room than you think you'll need—it sucks to accidentally bump into a wet painted panel!

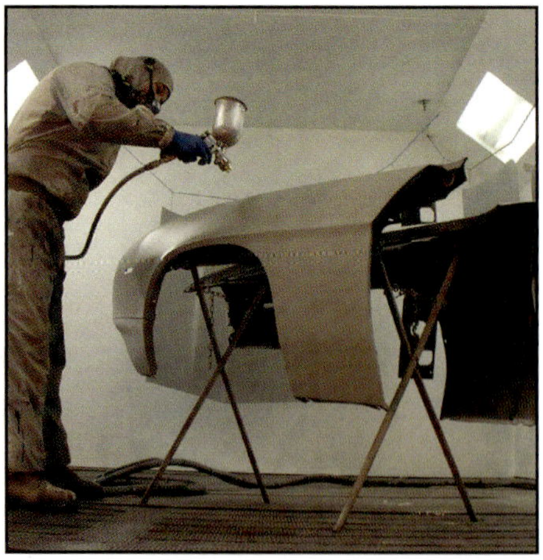

Mount Firmly
Another tip is to physically mount the body components to a stand (like these fenders are to this stand) or hanging wires. While we all want to believe we won't bump into a component, it does happen and the way to avoid a body panel falling to the floor is to take the time beforehand to mount it in position.

Here is the Mule body and panels just before it went up to the paint shop. I have it on a rolling stand (these are now available for purchase from Detroit Speed & Engineering, www.detroitspeed.com) that holds the body like it is sitting on the factory frame body mounts.

At this point, the Mule has all of its sheetmetal work complete and I am checking rough fitment of everything. Most of my cars are in bare sheetmetal at this point, but the Mule was E-coated, so the sheetmetal is black here. You can spray a car with an etching primer if you have to expose it to the elements during the shipment to the painter, but I prefer to let the painter start with a bare car so they can do their exact process to it.

The fastener on the left just came off the Mule and the one on the right has been cadmium plated. Amazing, huh?

Here's all the fasteners from the Mule after I got them back from the plater. Having all of the factory fasteners from a car stripped and plated usually costs me about $75.

The way I build cars is pretty basic: I am constantly installing/modifying/reinstalling parts and systems to make sure everything is going together like I want it to. Without fail, when I don't do this, I end up with something not fitting after being painted/plated/chromed. For this reason, everything goes on and comes off my cars at least three times.

Why I mockup parts: installing the trunk lid and torsion-type trunk lid springs really tries my patience and capability, but doing it shows me that all the brackets are properly placed (the body anchors for these springs were all cut out and welded back in during the mini-tubbing process—so this confirmed they were installed properly).

The Mike Verellen Paint Process

1. Receive car rough bolted together
2. Inspect for gaps in body panels, minor body damage, scratches, low spots, etc
   * *Note these in quote to owner and determine how these will be corrected*
3. Wash car with wax & grease remover
4. Fix minor body damage or issues
5. Disassemble vehicle body
6. Sand all surfaces with 320 grit smooth
7. Wipe down with wax & grease remover
8. Spray PPG DP40LF Etching Primer
9. Wait 12 hours, then spray on 3 or 4 coats of K38 high-build primer, letting each coat dry before applying another coat
10. Sand with 220 and 320 grit in a crosshatch pattern, spot fill low spots, sand smooth
11. Sand with D-A to remove sanding scratches
12. Clean surface with degreaser
13. Spray on more K38 Primer
14. Sand with longboards/etc to find/fill low spots
15. Sand with 220 grit sandpaper on D-A to remove scratches
16. Wipe down with wax & grease remover
17. Spray K36 low-build primer on surface
18. Block sand 'show' surfaces: 320, then 500, then 600
19. Wipe down with wax & grease remover
20. Paint 'cut in' areas like rollcage, trunk, interior, underside of car, etc
21. Spray primer/sealer DP48LF on entire car, let dry
22. Wipe down with tack rag
23. Spray color onto all 'show surfaces'
24. Wipe down with tack rag
25. Spray on multiple coats of Clearcoat, let dry
26. Wet sand with 600, 1000, 1200 & 1500 grit sandpaper
27. Buff clearcoated surfaces
28. Wipe down clearcoated surfaces with High-Gloss Polish
29. Rough assemble body, panels and swing metal
30. Shim all components to achieve smooth gaps and transitions

Here is another example of what you learn doing mockup work: the inner wheelwell opening for the front fenders did not clear the upper A-arms for the front suspension after I installed lowered front subframe bushings. To fix this, I had the sheetmetal cut back and formed at its edges to replicate the stock opening.

The sheetmetal that was formed on the fenderwells was left bare: the paint shop just sprayed this area with etching primer in the early steps of the process to protect it from surface rust.

There are two things that I really like about this shot: 1. Mike Verellen, owner of the paint shop and I are going over the quote and timing for the Mule paint job and 2. Mike's shop is clean and free of 'paint jail' queens!

Here are just some of the flat 'long boards' that Verellen uses during the surface prep stage. They vary in length, width and stiffness and are mostly built for specific situations.

The various roll boards and other sanding tools are all designed to speed and improve the result of the surface prep sanding. It seems to me like they do just that.

The first step for Verellen and his team was to inspect the bolted together car, note any issues, then promptly disassemble it and mount the body onto a custom-built rotisserie.

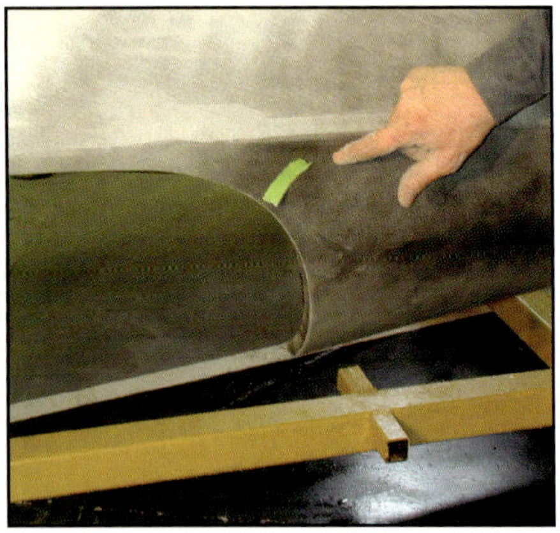

The Verellan team spends a lot of time doing an initial inspection of the vehicle when it shows up. He put small pieces of green tape everywhere he found an imperfection on the body (scratch, dent, rust, mismatched body panels, etc). Once he has had a look and stuck tape on the issues, he lists everything onto an 'issues' sheet so he can fix everything and account for the work.

Verellen is an innovator and here's one of his inventions: these 'longboards' he created and uses to speed the prep time of a body panel. Here, he is smoothing an area he identified as being 'low' and built up with some body filled.

The great thing about a body rotisserie is that it makes it easy to do any step of the painting process to any area on the body at an angle that is 'natural'—and if you are a painter, you know how important this is! This is physically and mentally challenging work and any advantage you can get while doing it pays massive dividends.

Here is Verellen after a few minutes of 'long boarding' the roof of the Mule. He makes a few strokes, then adds cross-hatch strokes at about a 45° angle on each side to reduce any scratches in the work.

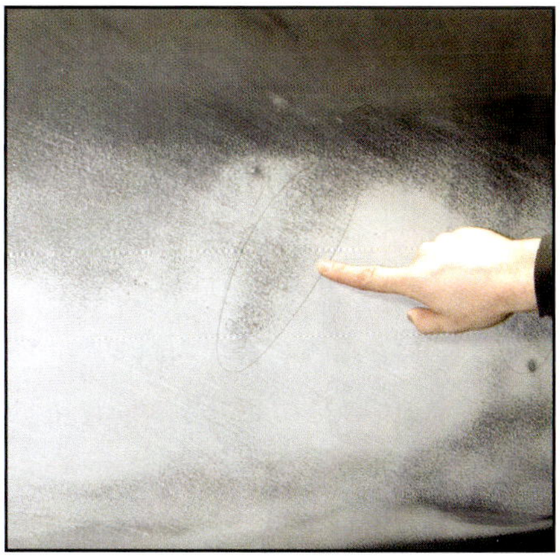

A 'low point' on the roof is quickly highlighted with the 'long board' sanding tool Verellen built. His team will add filler and sand it down on areas like this until the surface is mirror smooth.

The rotisserie also makes it easy to work on the floor of the car. This is the first car I had the floorpan fully prepped and painted, and the rotisserie Verellen built and uses makes it all possible. This is what the floor looked like when they started working on it (not bad, but not great).

I have used a Harwood fiberglass hood on a few of my builds because it fits tight to the cowl, is lightweight and looks good. Unfortunately, it also can hit the peak on the cowl panel. To fix this on the Mule, I had the peak 'flattened' by cutting the sheetmetal, hammering it down, MIG welding the section and grinding the welds.

The area was then bodyfilled and the fix sanded smooth. In all, this probably took 15 manhours and nobody would see it except me, but it's one of those things that just bugged me on previous cars and I decided to fix it on this one.

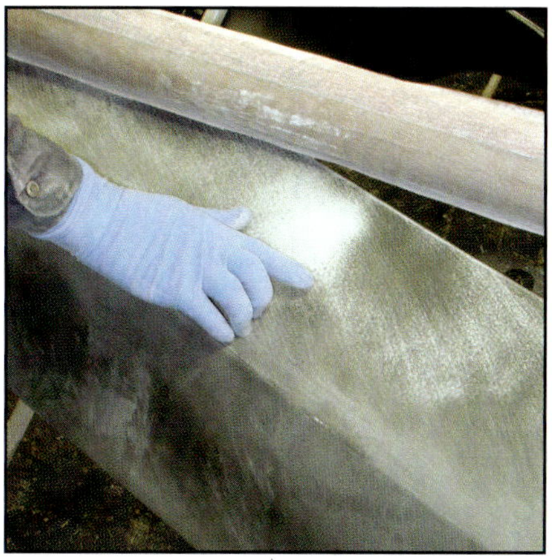

Here is a low spot (finger pointing) that has been highlighted by sanding the top of the passenger fender with a roll board. Notice the 45° angle cross hatch strokes that were made as the last step in the sanding procedure.

The Verellen team also has created these 'roll boards' to use on surfaces that aren't flat. It's the same concept as the flat long boards: put the sandpaper in contact with the high spots with a long, flat surface to make it easier to get to a 'mirror smooth' body.

Here are all the loose components I had E-coated. In general, the coating of these was a good thing as the chemical stripper was apparently neutralized on them and they were thoroughly coated inside and out (what I had originally hoped for).

The front subframe had been painted with DP 40LF, and after the paint cured, it was all sanded smooth before it was all painted again with high-build primer, allowed to dry, then sprayed with semi-gloss black paint.

Some panels are only going to need scuffing before being ready for priming/painting, like the front lower valence.

Minimizing the amount of dust and dirt that gets into the wet paint is a key part of ending up with a beautiful paint job. Here, the door has been hung in the paint booth and the inner door skin area is being wiped down with a PPG Degreaser-soaked rag before a coat of primer is sprayed.

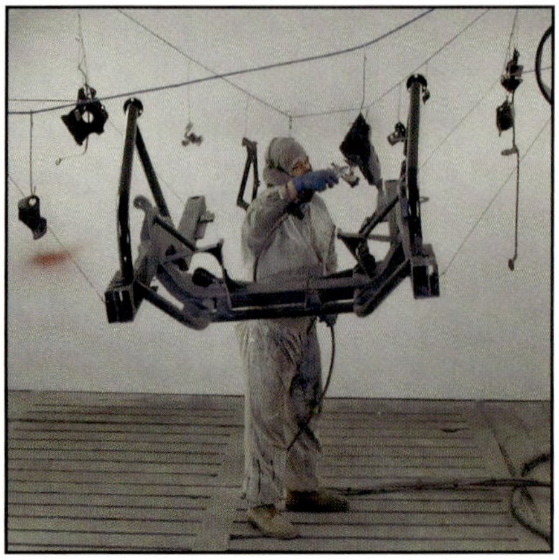

All of the parts that weren't powdercoated or plated needed to be prepped and painted, too. So each piece was stripped, then scuffed with 400 grit sandpaper, and then sprayed with PPG DP40LF sealer/filler paint in the spray booth.

The trick when spray painting parts is to hang them via holes that will leave the least noticeable 'witness' mark (the paint won't be perfect there but that area will be covered by another part or closeout panel).

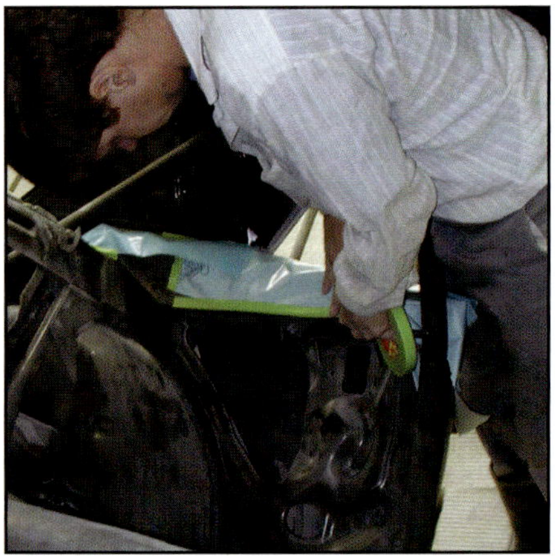

To avoid rollcage paint from getting onto the painted surfaces at the bottom of the rear quarter window, these areas were masked off. This takes time and costs money: ask your painter if they will be doing this when painting the interior/rollcage. It makes a difference in the end.

Yeah, painting around rollcage bars inside an interior is not pretty, in general, but with the car on a rotisserie, the painter can have an assistant move the body around to make it as easy as possible—like here where the painter kneels in the door opening while spraying.

Getting a smooth, even coating of paint on rollcage bars requires the painter spray on paint very judiciously, moving the paint gun constantly while knowing when to apply a light stream of air on the surface to 'flow' the paint out.

For me the paint customer, I love to see a mixing area like this: clean, organized, well stocked with premium PPG paint and ready for use. This is a tough area for painters to keep clean as it is heavily worked, but it is indicative of their attention to detail.

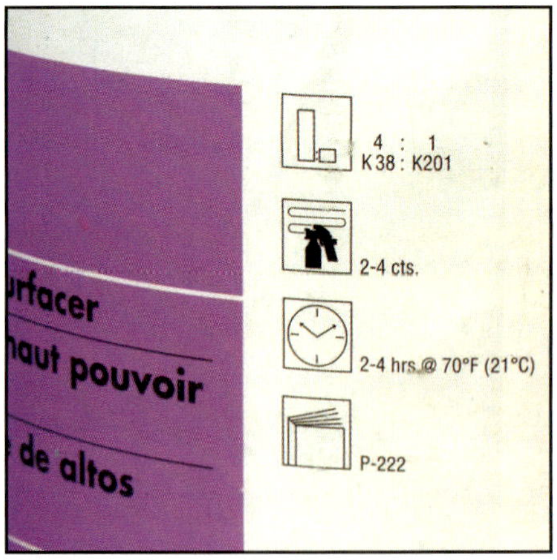

The PPG paint products all have these image-style instructions on them showing how to mix and apply the paint. Interestingly, every painter I have ever worked with used their own 'mixtures' they developed for mixing and applying paint products—and they usually had a favorite paint company.

I recommend working with a painter with a 'downdraft' paint booth that is drawing air through a wall of fresh filters (as shown) to minimize the amount of dust and particles that end up in the wet paint...which helps the surface looks its best and minimizes the amount of sanding required afterwards to smooth the surface.

With the floor of the main body prepped/primed, the focus of the work became getting the 'swing metal' panels (like the doors, hood, trunk) 'straight' and ready for final paint. That task looks like a lot of sanding and filling, sanding and filling...

The front fenders (in the background), also fall into this mix as they need to be perfectly straight and line up with the hood, front fascia and doors with all gaps and surfaces worked to fit properly and look great.

This is what the tedious work being done at this point is all about: sand an area, spray on a light 'guide coat' of black paint, then sand it again, fill the 'low spots' (in this case the dark colored area), let dry, sand again until all surface is one color after light sanding.

E-Coat Misfire

By far my least happy moment building the Mule was when I had its body E-coated (by a company that is no longer in business).

Actually, the E-coating process did exactly what I had hoped it would do: coat all the sheetmetal with a protective coating to help it last a long time. The problem with the process came from the chemical paint stripping that was required to be able to apply the E-coat—while the stripping company told me the stripping chemical would be neutralized before the E-coat was applied, it obviously wasn't as we discovered the doors rusted through at the edges in less than a year on the road (the chemical got between the door skin/door structure and the neutralizing agent did not...). We replaced that rusty sheetmetal, painted it all up and moved on.

So what did I learn? Well, that when looking at a new technology involving chemicals, I didn't ask enough questions of independent third parties. I only spoke with companies trying to sell me a service: now, I don't blame those people for my choices.

I made them and accept the outcome. But now I am even more vigilant in understanding EXACTLY what is happening with a process before I greenlight it on one of my cars.

Fender Fixing

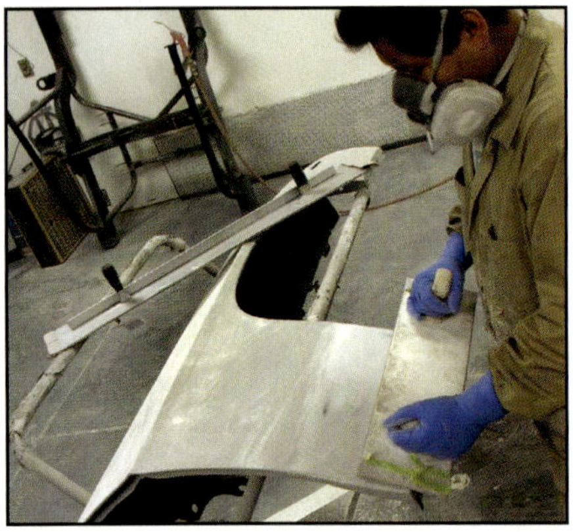

There are no shortcuts to getting the body 'straight'—it involves lots and lots of filling/sanding/filling/sanding. Here is a good example of the work and tools needed to get the surface ready for painting.

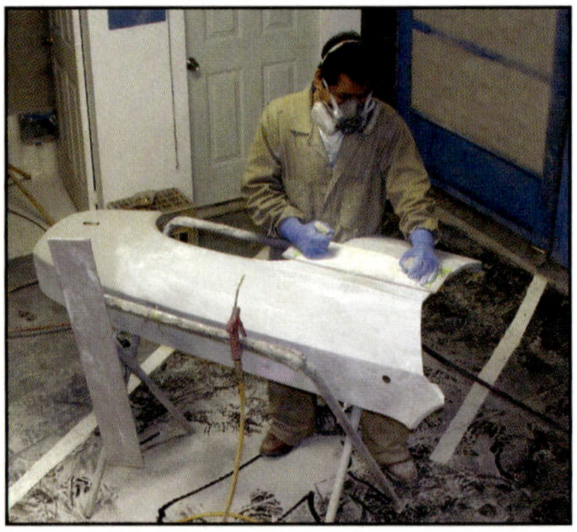

The Mule's front fenders are like most body panels that need to be smoothed: some flat surfaces with lots of peaks and curves. You will need to have and know when to use many different sandpaper backing shapes. Here, a flat backing tool is being used with 180, 240, 320, then 400 grit sandpaper on a panel that has been sprayed with PPG K38 High-Build Primer.

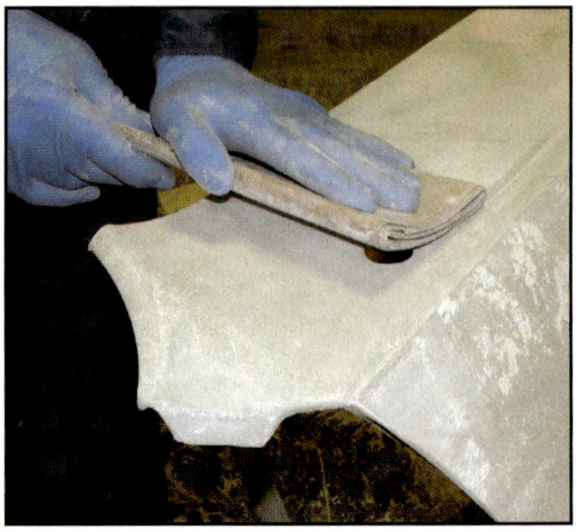

Each area of a body panel needs to be smoothed with a grit sandpaper before the next higher grit (the higher the number, the less abrasive the sandpaper) is used. And curved areas need a curved backing (like the 'flat roll' shown) to mate properly with the surface.

A 'long roll' is used to smooth curved areas that have a long radius in them, like the top of '69 Camaro fenders. This keeps the sandpaper flat enough to highlight low spots in the sheetmetal while not digging into the curved surface.

The long flat sandpaper backer works on this one edge of the Mule's front fender. The sandpaper is stroked longways…up and back. Then scrubbed in short strokes sideways and then diagonally. This way, the surface is sanded in multiple directions. This is done for every portion of the painted surface.

Low spots are what the sanding is highlighting (circled with pencil). By watching where the sandpaper is not touching (easier if you spray a light 'guide coat' of paint on the area you are sanding), you can see where to add spot putty filler.

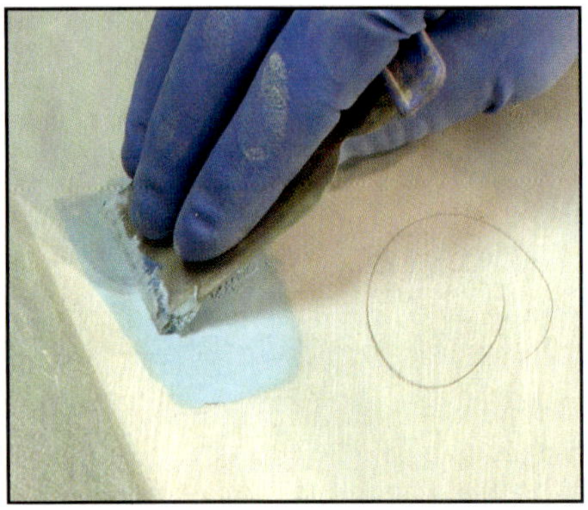

This is what it looks like to add spot putty filler. It is mixed as needed 50/50 putty/activator, then 'skived' onto the surface with a small putty knife. It needs to dry about 30 minutes, then can be sanded smooth. This process is repeated until an entire body panel is smooth.

Smoothing the Firewall

We all like the idea of having a 'smooth' firewall, but actually doing the work (or having it done) is something all together different. This is a quick lesson in how to smooth the firewall for maximum improvement with minimal investment.

The key to smoothing the firewall but keeping costs/time to a minimum is to only add filler where you plan on sanding—like the area around this bolt hole at the top of the firewall (near the mating point for the Detroit Speed Firewall Blockoff plate).

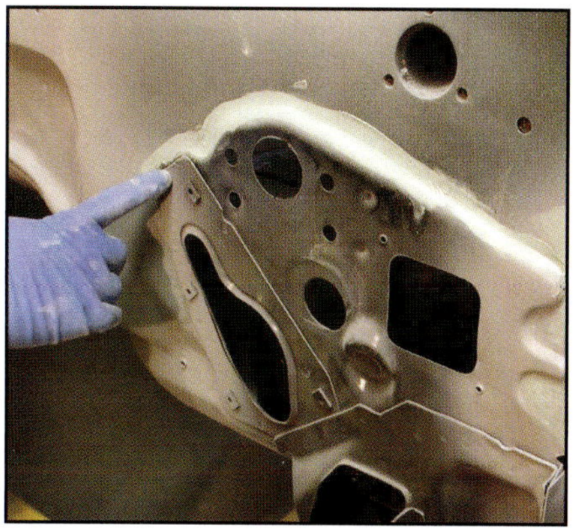

On this area where the steering column/brake booster/fuse block/etc come through the firewall, only the top edge of the firewall is getting filler—the rest you can't see behind components and would take FOREVER to sand smooth anyway!

Final Prep

After everything has been sanded flat and smooth, all of the surfaces receive one last sanding with 420 grit sandpaper while rubbing the free hand over everything to check that the surface is smooth and free of imperfections. Any issues are corrected with filler/sanding and the body blown off/washed off with Degreaser in prep for painting color.

The key to having a paint booth do its job is to keep any dust out of it…which is really tough to do in a dusty body shop! For this reason, all of the sanding work on the Mule was done in one area, then everything completely washed down before being moved into the paintbooth (shown)—even the rotisserie and its tires were cleaned!

Applying Final Primer

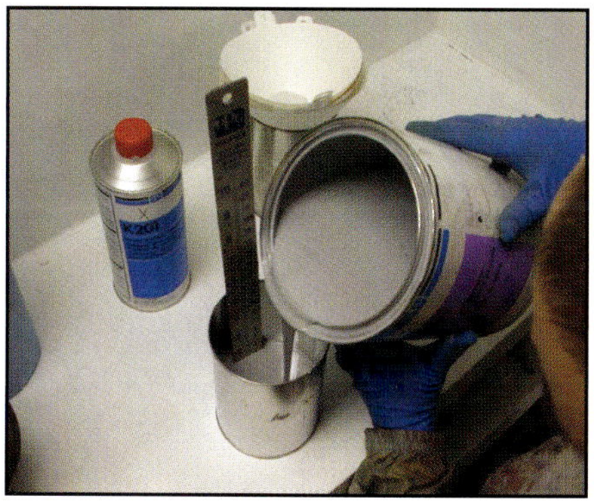

The PPG K36 Primer is mixed 4:1 parts of paint to activator. Mixing up about a quart mixture will cover an entire '69 Camaro car body and a few body panels and will last about 1 hour before it will have 'kicked' for too long and should be discarded for a new batch.

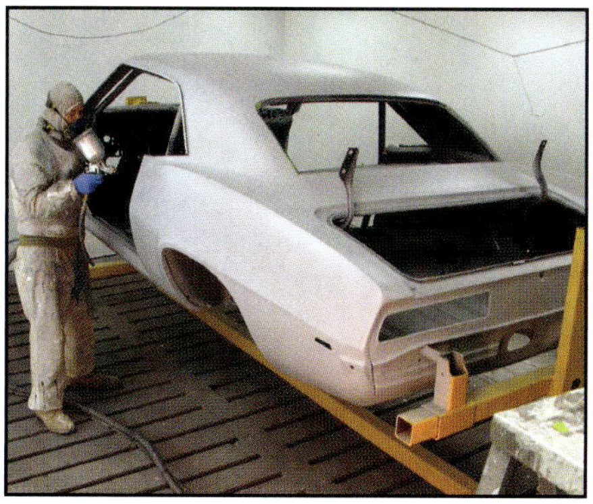

The PPG K36 Primer is applied in a paint booth with supplied-air to the painter, as the fumes are rather aggressive. The surface to be painted should be wiped down with Degreaser before the painting begins and paint should be applied with a High Volume, Low Pressure (HVLP) spray gun (room temperature air).

The bottom of the Mule was sprayed with K36 Primer in a few select areas that were bodyworked to seal the filled areas in prep final painting them.

Prepping the Bottom of the Car

I have never had the bottom of my car 'bodyworked' before the Mule, so this was new to me, but it did look amazing when complete. The process involved sanding and filling then sealing the seams where the sheetmetal mates, before applying paint intended for the abusive nature of being under the car.

Verellen uses these small DA sanders to get all the nooks and crannies sanded smooth in a short amount of time. He went through 180, 240 and 320 grit sandpaper in prepping the surface.

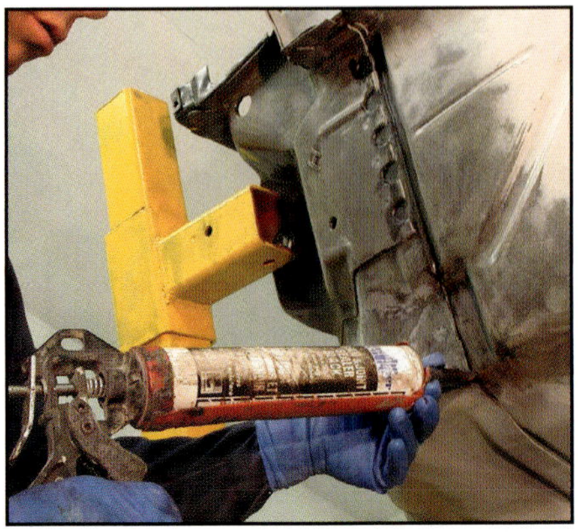

The seams where the various sheetmetal panels mate together were then sealed with seamsealer. This process involves pumping the sealer onto the seam with a handheld applicator…

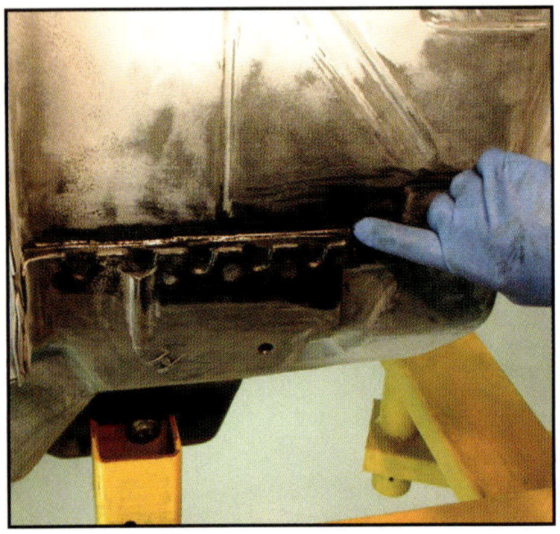

Then, the sealer is 'wiped' smooth on the seam with the finger in a gloved hand that is wiped on a paint thinner-soaked rag. This minimizes dust/noise/exhaust/etc from leaking into the interior and also makes the bottom of the car look way better than stock.

In prep for painting the bottom of the car, the body was rolled into the paint booth again and the entire body covered with protective plastic (taped in position). The floor was then wiped down with Degreaser in prep for spraying on primer/sealer and semi-gloss black epoxy paint.

Painting the floor on a car is soooooo much easier when the body is able to rotate on a rotisserie. This is when they are really worth their weight in gold—and the results are where it all shows up.

Taping Off

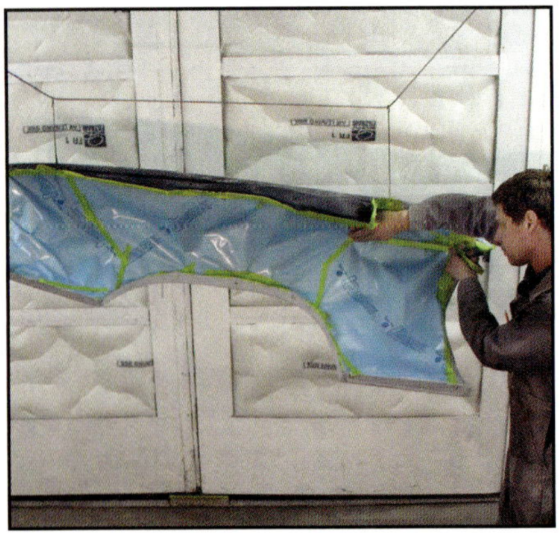

In preparation for spraying with color, all the previously painted surfaces (like this fender inner which was painted semi-gloss black) were masked off with Bonmask paper and 1 inch green masking tape.

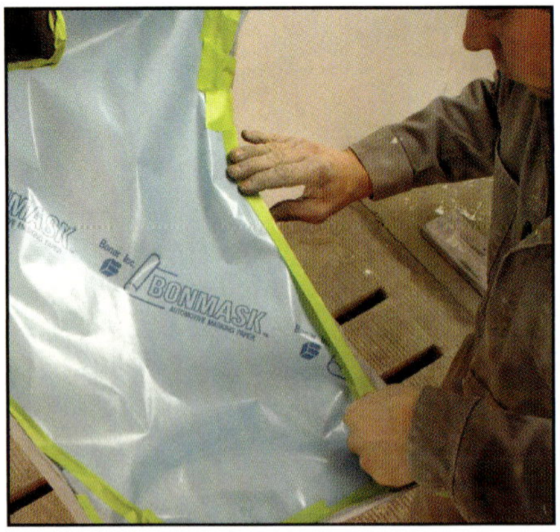

Using Bonmask and real masking tape adds to the cost of the paint job, but once you've come this far with a paint job like this, cutting a few dollars out at the expense of shoddy overspray on panels is not an option.

It is always a good idea to tape off any area you don't want to be painted, as modern paint is very good at adhering to things! For this reason, the openings to the door inners were masked over along with the door latch hole.

Blowing Off/Wiping Down...Again, Then Applying Sealer & Basecoat

With all the components masked off and hanging in the paint booth, the entire booth was blown off with compressed air and the floors washed down with water. The parts were then wiped down with Degreasor.

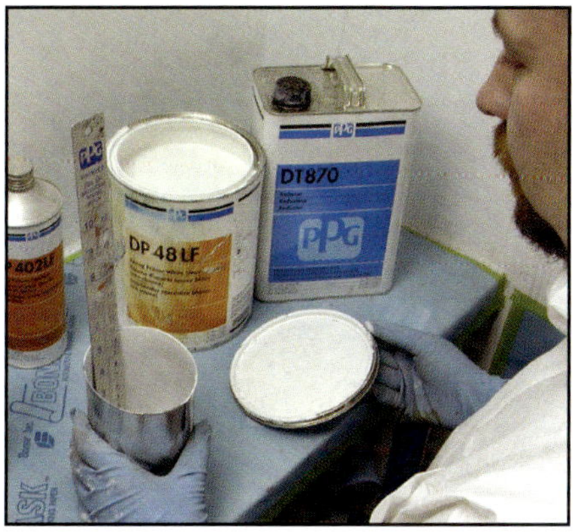

The paint chosen for the Mule was a silver that Verellen decided needed a white basecoat to properly 'pop'. Here is the white epoxy sealer being mixed up before being sprayed on all the 'show' surfaces of the body. This paint is mixed 3:1:1 paint/catalyst/thinner and needs to be sprayed on the car once the base coat is tack dry.

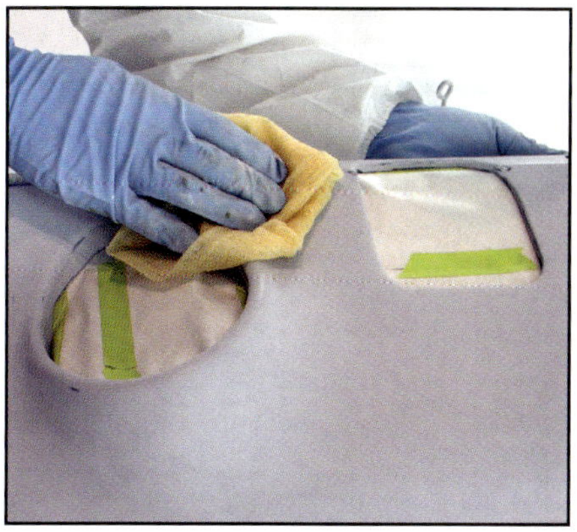

Just before spraying on the primer/sealer, the surface to be painted was wiped down with a fresh tack rag to remove and debris that had settled on it.

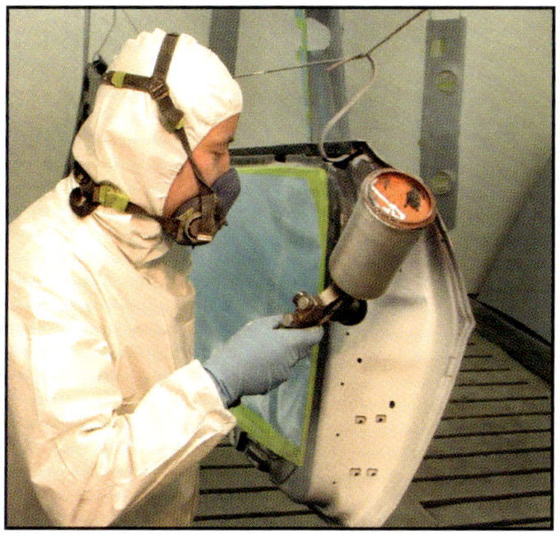

The white-tinted PPG DP48LF primer/sealer is sprayed on to prevent any seepage of the bodyfiller or primer into the color basecoat. After this sealer 'tacks off' (which means if you touched it, it would slightly grip to your skin), the surface is ready for being sprayed with color paint.

The PPG color basecoat paint was mixed 1:1 with reducer, then poured into the HVLP gun and sprayed on the body panels of the Mule.

Once the body panels were dry, they were removed from the paint booth and the main body rolled in. The body was fully taped off, wiped down with Degreasor and a tack rag, then sprayed with Primer Sealer.

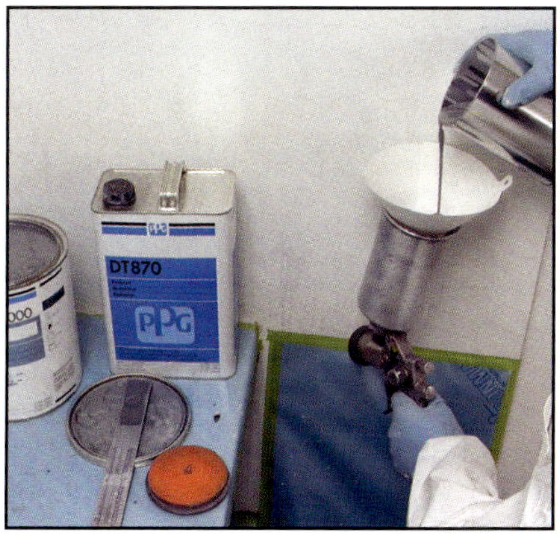

While the white Primer sealer was 'tacking up', another batch of color basecoat was mixed up in a 1:1 ratio of basecoat paint to reducer.

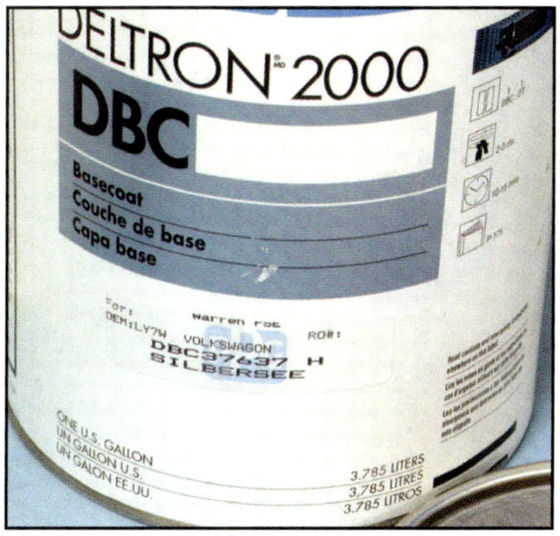

The Mule required about 1 ½ gallons of PPG basecoat to cover all of the 'show surfaces', which is not unusual for a vehicle of its size.

Applying Clearcoat

Between the spraying of the basecoat and clearcoat, the taped off areas were checked to make sure everything was staying in place. This is the attention to detail that makes a paint job really standout.

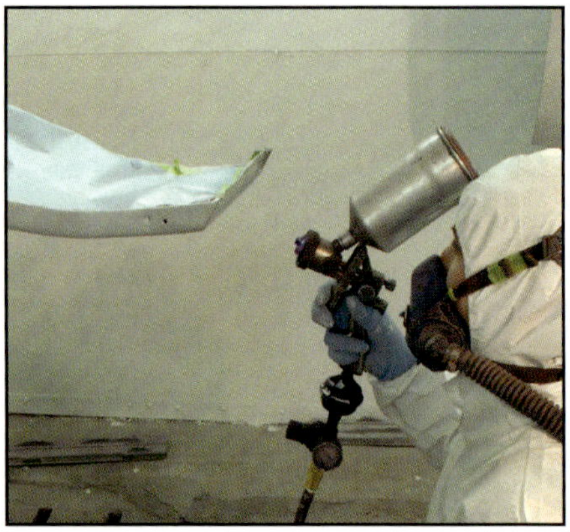

The PPG Clearcoat DCU2002 was mixed up and sprayed onto the Mule in multiple coats. Mike Verellen recommends spraying on at least one more coat of clear than you think you'll need—this just gives a little more thickness to sand off during the colorsanding process.

Colorsanding

While it might seem like you are 'done' when the basecoat/clearcoat goes on clean, the pros know there is more work to do: namely, colorsanding. This tedious, finesse-laden step is what takes all the work to this point and really makes it shine.

The first step in preparing to colorsand a body panel is to protect the peaks and hole openings. On the Mule, this was done by applying ¼ inch wide 3M 233+ green, all-weather tape on the peaks and 1/8 inch wide 3M 233+ tape on the hole openings.

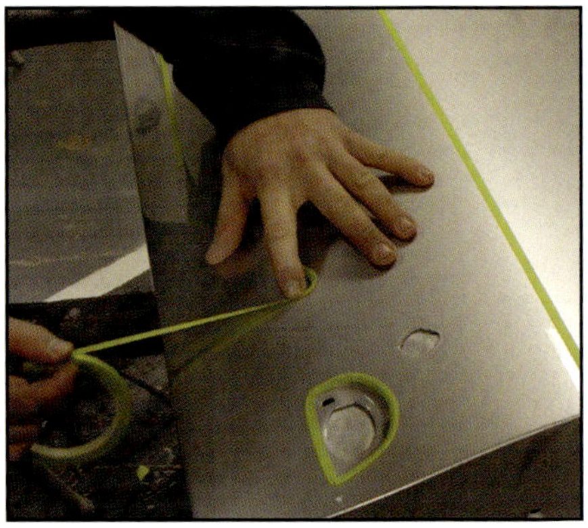

The 1/8 inch wide 3M 233+ tape is used on the hole openings because it is narrow enough to be 'swept' around the circumference of the hole in one strip—the ¼ inch wide tape bunches up if you try to do it on the 'tight sweeps'.

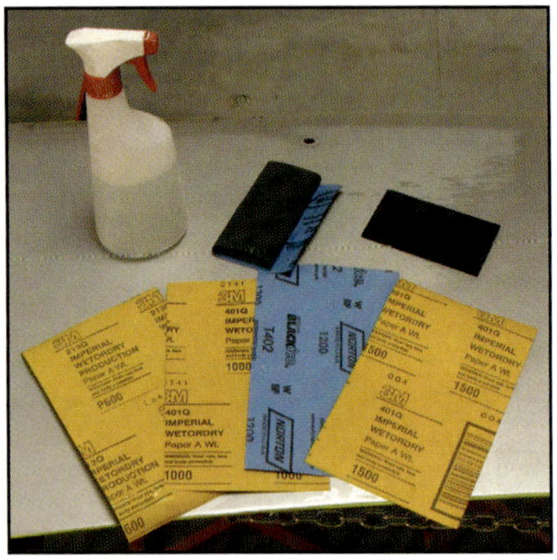

This is what Verellen and his team use when colorsanding: a spritz bottle filled with water, a rubber backer, a water squeegee and wet sanding paper that varies between 600 up to 1500 grit.

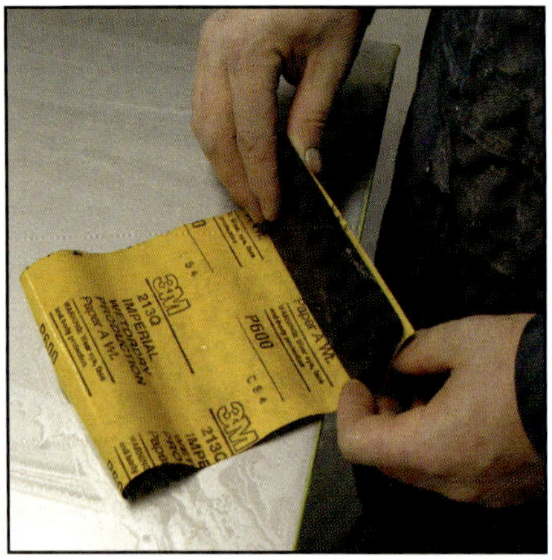

Verellen has found the best way to colorsand is to start by rolling a section of sandpaper up on a rubber backer.

A spray of water is applied to the small area being sanded and the light sanding begun. Water is constantly reapplied, but in light amounts, to help 'flow away' the material being sanded off the surface.

Colorsanding (cont.)

With the sanding pad cradled in the fingers, Verellen and his team would sand back/forth, side/side, then at 45 degree angles in one area until it was smooth. Then, they'd move over to the adjacent area and start over.

To see if the surface is 'smooth', Verellen and his team use a small squeegee to wipe water away. The area that has not been sanded yet will look like the surface of an orange (thus the nickname 'orange peel' for unsanded clearcoat) when wiped dry.

The universally dark areas on this wetsanded area are considered 'smooth', while the shiny areas need some more sanding. The team knows when to switch to less grit sandpaper (like go from 600 to 1000 grit) when all they see are more scratches.

A large chamois is used occasionally over the entire panel being wetsanded to wipe away any water and material on the surface. This just helps to prevent the panel from getting covered in goop that makes it hard to see how the wetsanding process is progressing.

When wetsanding close to the edge of a panel, a VERY light pressure is used to avoid 'blowing through' the clearcoat and basecoat. This is where the pros show their hand as things can go wrong very quickly in these areas…

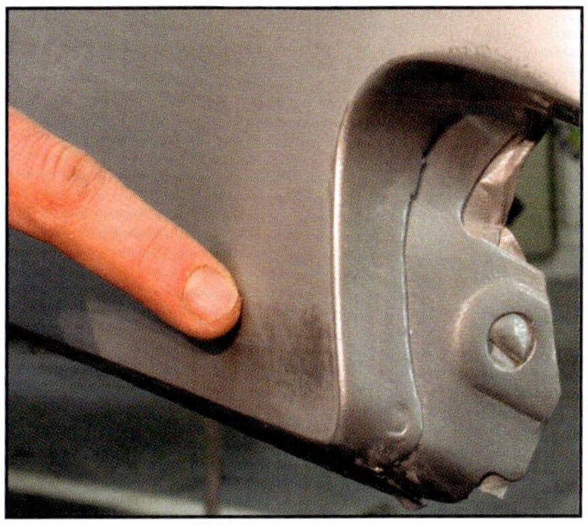

This is a good example: there was a low spot right down near the lower edge of the panel, so the sanding would have gone through the clear/basecoat if not for some skill.

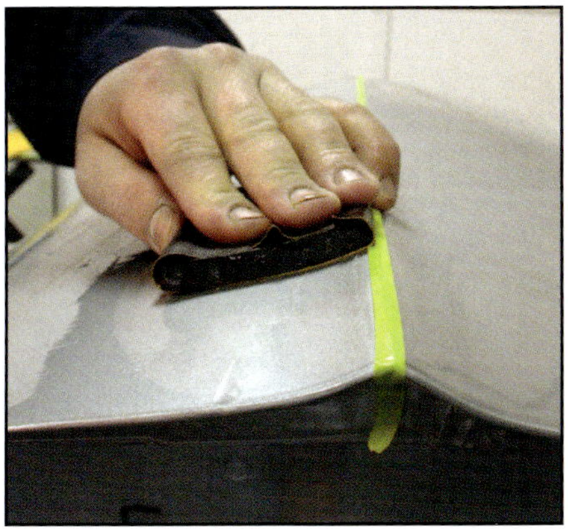

Extreme caution is given when approaching high spots (like this 'peak' in the door). The 3M tape does a lot to protect from 'blowing through' the clearcoat into the basecoat, but Verellen doesn't rely on the tape—he is very cautious around these areas.

The area around these holes is going to be under the spoiler that is mounted on the trunk, so Verellen did little wetsanding here: why expend time and effort sanding an area that nobody will see?

This area has been squeegee'd dry: the light area on the left has been sanded smooth with 1400 grit wetsanding, while the area on the right is unprepped. As you can tell, it is easy to see where the wetsanding has occurred with this trick…

Colorsanding (cont.)

Once the entire panel has been wetsanded smooth, it is wiped down a final time with a wet chamois. As a tip, the chamois is cleaned out by dipping it in a bucket of clean water and squeezing it out a few times. Squeeze it out before wiping down the panel again.

Buffing It Smooth

Once the clearcoat has been sanded smooth, the buffing starts. Verellen uses Meguiars buffing/polishing products with a variable speed buffer, foam cutting- and polishing pads, a squirt bottle and a chamois. Here, some buffing compound is being applied.

To remove the wetsanding scratches, a 3M Compound pad (pn #05723) was spun at about 1,600 rpm with Meguiars universal compound (pn SPC2). Then, to remove the pad swirls, a 3M Polishing pad (pn 05725) was spun at 1,200 rpm with Meguiars finishing glaze (pn SPC10) until the swirls are gone.

When everything looked beautiful, it was all bolted back together. I just love this underhood shot because it what makes Verellen does so unbelievable. This is the bottom of the Harwood fiberglass hood after they were done with it: this is amazing work!

Here's another example of how incredible the Mule looked when it left Verellen's shop. It is tough as hell to prep and paint parts to look great, but then to get them to go back together cleanly and look great is even more tough.

By the way, this is the cowl panel after it was fully body prepped/painted: you can't even tell we 'dropped' the front tip of the peak in the center of it (to clear the Harwood hood). Nice, huh?

With all the paintwork complete and the body reassembled, Verellen put it back on my rolling stand and brought it back to my shop. Here we are rolling it out of the trailer so I can get to work putting it together. If he looks happy, its because the I am happy with the way the car looks and he can has one more car 'done'!

Hardcoating Suspension Components

Getting it all looking great but making it durable, too

While building and driving cars is what I really enjoy, I also like to have a ride that looks really good—but I am not willing to constantly rub/scrub/clean/polish to keep a car looking new. This desire has kept me looking for ways to build a car that requires very little maintenance…and led me to 'hardcoat' anodize the forged aluminum suspension components on the Mule.

I really like the way the Mule suspension parts looked with the hardcoat and have used this process on many components since my initial experience with great results. I can say the ability of hardcoating to resist chipping or pitting is truly impressive—a lot of the Mule suspension parts look brand new after many years and many abusive miles.

I worked with a Detroit-based company, International Hardcoat, Inc, and was really impressed with their abilities and the appearance of their hardcoating.

To me, hardcoating should be considered along with powdercoating, chroming and painting when deciding how you are going to prepare your aluminum components.

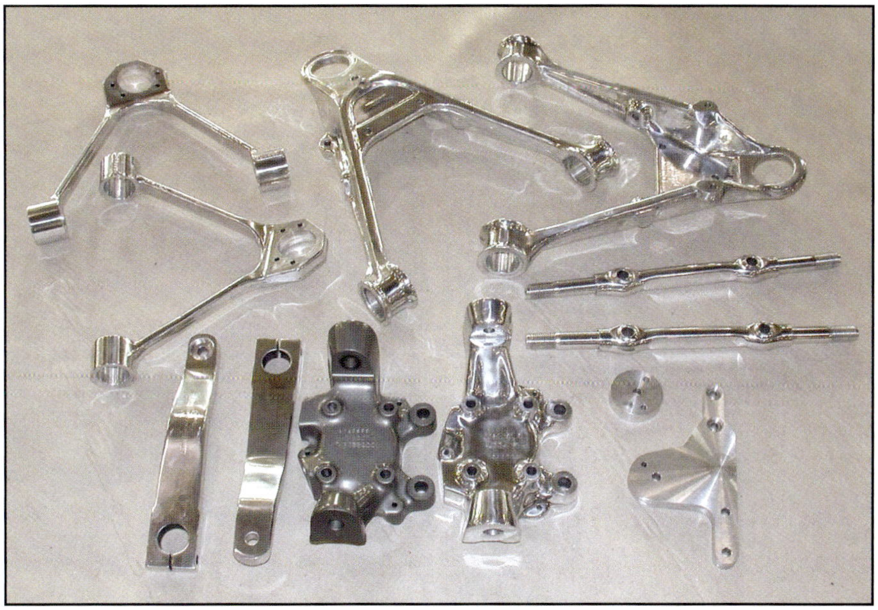

While these polished aluminum suspension components look great straight from the polisher, we all know they won't look that way for long on the street—which is why they are going to be hardcoated and look like the one gray upright in the center of the photo!

Coating Steel

There are many different ways to protect steel or iron parts: like chroming, powdercoating or painting. On the Mule, I used all three, plus I decided to nickel-plate the Speedway Engineering swaybar because I like the way the nickel plating looks when it is new. Also, nickel will undertake a slow discoloration process that I think looks good…and different, too. To me, choosing different colors and treatments for the various components on the car is one of the really creative parts of a vehicle buildup.

Hardcoat Anodizing

Anodizing is an excellent way to minimize corrosion and pitting to the surface appearance of aluminum. Hardcoating is considered a Level III anodizing process and can be performed on polished, machined or rough-finished aluminum.

It is created using an electrolytic solution of sulfuric acid held at a specific temperature and charged with an electric current. The parts are held in the solution for between 20 minutes and two hours depending on the material being coated and the desired coating thickness. Unless a dye is added, hardcoating will produce a gray-ish coating that is 10 to 50 microns thick with 50% buildup and 50% penetration into the parent material. More parts than you can think of are hardcoated and for good reason—it is an excellent surface prep.

To start the hardcoating process, each part is completely cleaned of any debris. Most of what is removed is polishing compound, which gets jammed into every crevice and hole on a part during the polishing process.

Next, each hole that shouldn't have hardcoating in it is plugged with these hardcoating resistance stoppers. If you don't do this on holes where a tight fitting bolt or swaybar is supposed to engage, you will be drilling/reaming those holes out later...

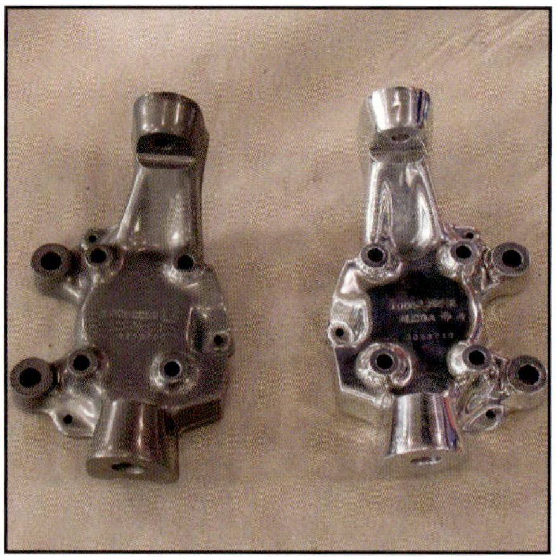

While I love the way aluminum looks freshly polished (on right), I really hate polishing it every week—which is why hardcoating (part on left) is so appealing to me: you get a great looking part but none of the maintenance issues.

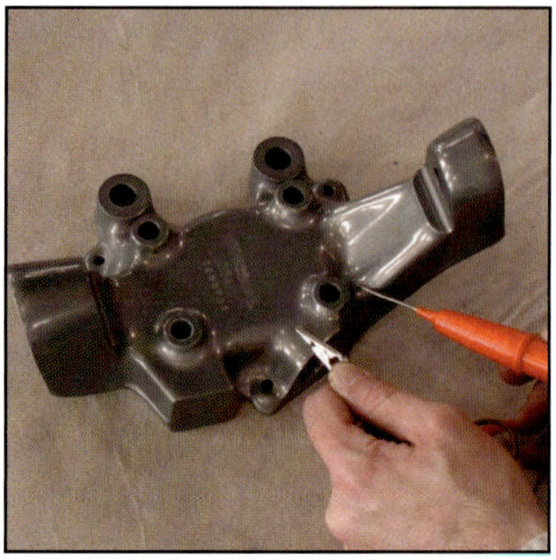

Each part I brought to be hardcoated was checked by the IHC team to make sure it had similar conductivity. They did, which meant they could be coated together.

These extruded aluminum components give you an example of the colors you can hardcoat onto parts.

I have always liked the way CV Products components looked, so I took this oil filter mount over to International Hardcoating and told them to copy the color...but a little darker.

Just so you know, if you hardcoat a polished cast aluminum part and a machined aluminum part, the hardcoating will look slightly different. This is because the final appearance of the hardcoat depends heavily on how surface prep: just an FYI.

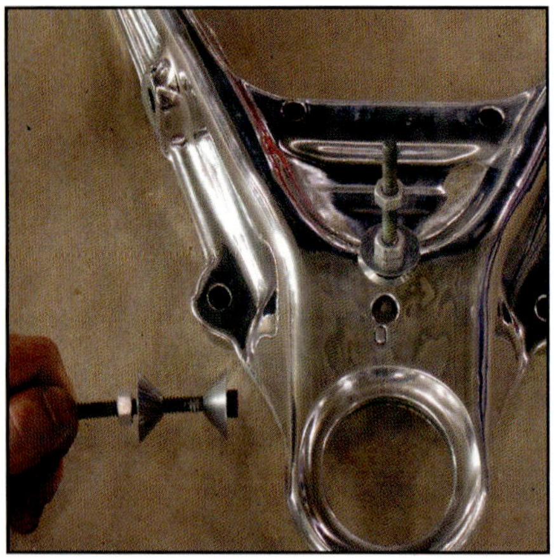

To fully coat the entire surface of a part, these titanium (Ti) bolts/nuts and tapered shims are used to 'grab' a part in a low visibility area and touch as little of the 'show surface' as possible. Ti is used as it is most resistant to the corrosive nature of the hardcoating process.

The parts were then bolted to these Ti racks (the use of the expensive Ti material is required as it is highly resistant to the caustic hardcoating process) in preparation for being hardcoated.

These are the tanks where the hardcoating process takes place. The racks, with the parts on them, are moved from tank to tank via computerized hoists that follow a proprietary sequence of chemicals that clean the surface, etch it to prepare for the hardcoating, then create the hardcoat on the part.

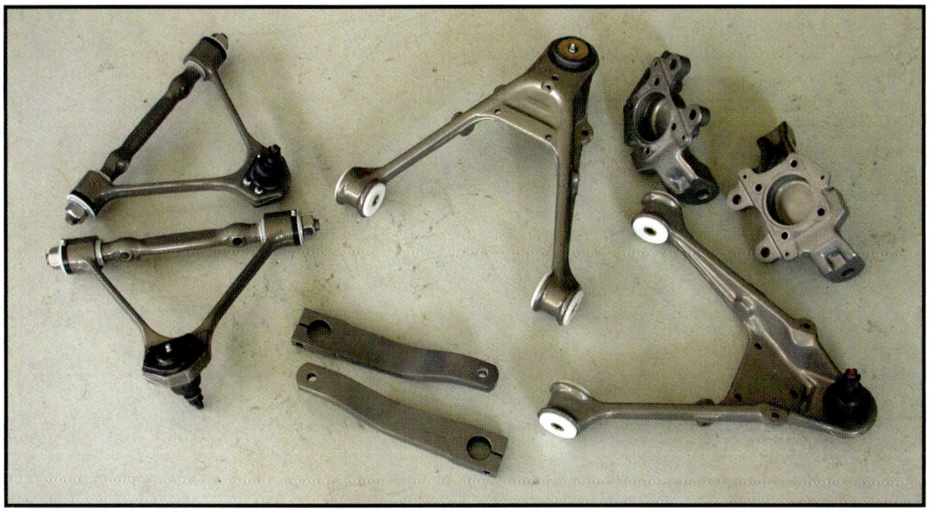

This is what the suspension parts looked like after completing the IHC hardcoating process. To me, these parts looked amazingly cool the day I bolted them on and, even better, they look pretty much like this today…almost 10 years later!

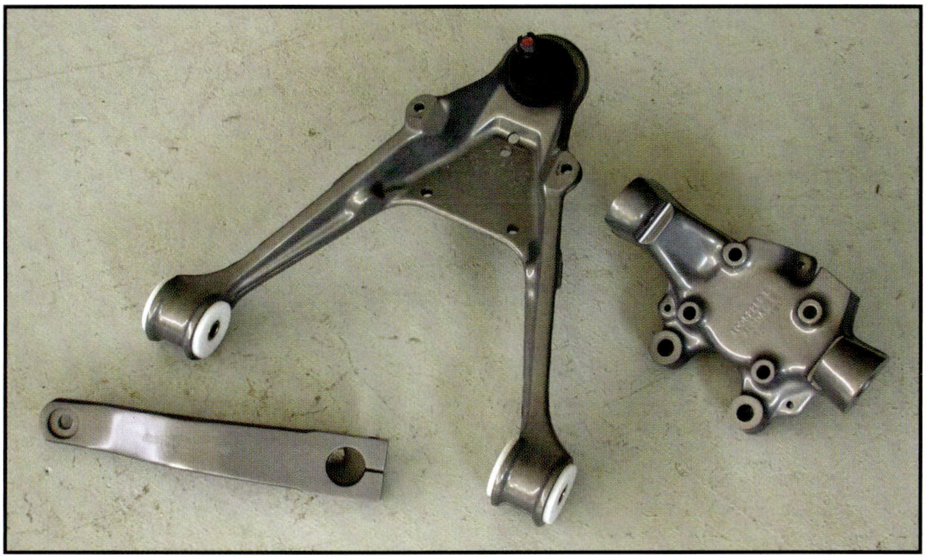

I do recommend you prep the surface of all the materials to the same finish. For the Mule, I had everything polished to a high luster (even the machined Speedway Engineering swaybar arm).

Another idea is to combine polished and machined surfaces, like this Vintage Air 4-hole bulkhead fitting. I did this one part on the Mule and liked the way it looked—I will probably do more of this on a future buildup!

The Post Paint Buildup

Gaining speed to the finish

I really like it when I get a car back from the paint shop as it is a clear sign we are making progress. For me, it probably is only the halfway point in the project, but it feels like it is a lot further along…I can see it as a completed project!

I play a little game with myself when I get the painted body back: I bolt on just enough pieces so it looks like a car. This gives me additional energy to attack the project and keep moving forward!

Here is the Mule's body the day I got it back from paint shop. I usually clean it as well as possible, then roll it into the shop and start bolting on parts.

Here's the Mule body after about 2 hours in the shop. You can see I have mocked up a headlight and I'm just getting ready to bolt on the front grille. This helps it look like a 'car' and keeps me moving forward!

Seal Tuneup

Doing the details on the rubber seals

It's true I like to build cars that look cool and are really fast, but I also like my cars to look and work as real cars. One of my real beefs is when the rubber seals either don't fit properly or don't work properly. So here is what I do to install my rubber seals.

The trunk seal is probably one of the easiest seals to install—but with that said, there is still a process to follow when installing one. I first measure out the midpoint of the trunk opening at the front, then put a small mark down in the trough with a Sharpie pen.

I then lay the seal in the groove all the way around the trunk to determine what the length it should be—starting the seal at the exact centerpoint of the trunk opening. Then I mark the bottom of the rubber seal with a marker. This is not how the factory does it - they put the joint at the bottom of the trunk opening but I don't like the way that looks. This might let a few drips of water in during a heavy rain but I think it looks a lot better and will put up with it.

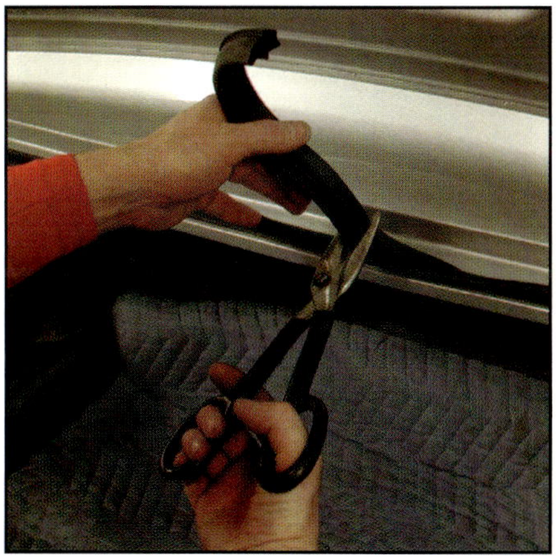

I then line up my large scissors perpendicular to the length of the seal and cut through it in one fell swoop so I get a clean cut. This sounds easy, but the seal is pretty stiff, so it takes a lot of force!

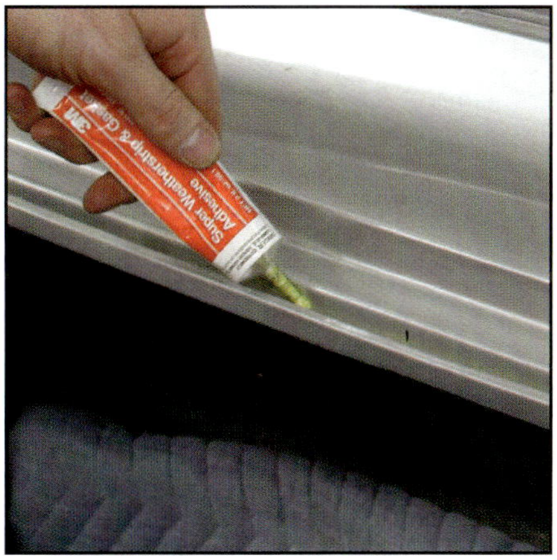

I then remove the seal and start to lay a 1/8 inch bead of 3M Super Weatherstrip and Gasket Adhesive glue in the seal trough.

After completing the apply of the glue, I let the glue 'tack up' before installing the seal into the trunk seal trough. The install starts with one end at the centerpoint of the trunk, then I continue laying the seal into the trough around the circumference of the trunk until the ends mate up

Door Handle Seal Tuneup

I like to use stock door handles on my cars (because they actually work), but getting them to look their best with reproduction seals takes a little detailwork. So everything looks and works great, I order repop door handles (and just about everything else I can get reproduction) from Year One.

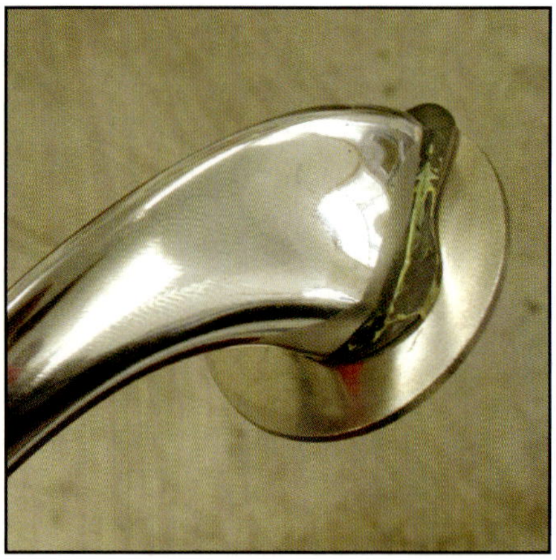

On the anchor portion of the door handle, I apply 3M glue to the handle-side of the seal, then sandwich it in place with a large washer and the door handle bolt—which invariably pushes the seal out from under the handle…like this.

Using a sharp X-acto knife, I trim away the portion of the seal that is protruding out from under the handle. I leave the handle/seal bolted together for a few hours, then when I remove the bolt, the seal is firmly mated to the handle.

I do the same on the button portion of the handle: sandwich seal under handle and trim away any excess seal material.

I also do the seal tuneup on the key cylinder by gluing the seal to the cylinder, then installing the cylinder on a piece of metal using the tension spring. This type of detail work is what makes the car look 'right'.

Front Suspension Final Build

What to do on the 'final' assembly of the front suspension

You've seen me do the initial mockups to get the front suspension system working, along with some modifications to improve and personalize it and then the process I went through to get the parts looking good. Now, I'll show you what I do to get everything back together, looking good and ready to run for the long haul/on-track beatings.

Important!

The steering and braking systems on your Pro Touring car are VERY important to both safety and performance. This point in the build is the last time before the car hits the road where you'll be able to check whether they are up to the task of setting the direction for the vehicle and getting it whoa'd up.

Use this moment to make sure the steering works through the entire suspension travel the vehicle will have (this can be done by installing the shock absorbers without the springs and using a floor jack to move the suspension through its entire travel while turning the wheels). Also, make sure the brake lines are long enough that when the suspension is at the outer reaches of its travel, the hose is not about to be pulled off the frame. Once you get the wheels/tires mounted up, run through the whole carnival of pain again to check that everything clears everything else and can do its job. These are the kinds of checks I do during the final assembly to make the chances good that my initial road miles will be fun and problem-free.

For more on getting a front suspension together and making it work, check out the following sidebars and photos/captions.

The Mule front suspension is a work of art because I spent the time to get it all fitting/working 'right', got it all polished/coated/prepped and then spent the time to get it installed properly. This is one of the nicest looking front suspensions I have ever built and here is how I do the final steps so you can have a suspension that looks this good!

A-arm Assembly Process

There's a reason I get pre-assembled suspensions from Detroit Speed nowadays…

I don't do a lot of the tricky work enough to be able to just power through it. One of those jobs is assembling bushings into front A-arms. It requires building special tools, using a powerful press and a lot of 'touch' when making it all come together. So, for those of you considering doing this type of work or just wanting to understand how it's done, here is the process I used to follow when assembling front A-arms.

1. *Organize bushings in prep for assembly*
2. *Apply oil to surface of bushing being pressed into A-arm*
3. *Press one pivot shaft bushing into A-arm*
4. *Install shaft on installed bushing/A-arm*
5. *Install 'spreader' between two legs of A-arm*
6. *Press second pivot shaft bushing into A-arm*
7. *Press/install ball joint into A-arm*
8. *Pump grease into zerk fittings until it starts coming out of joints*
9. *Install A-arms on vehicle*

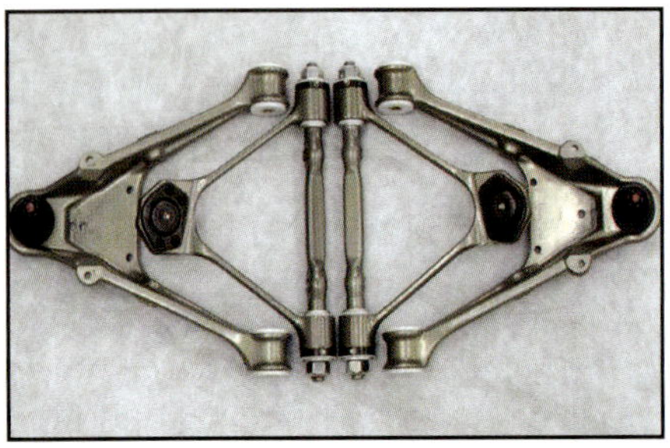

Assembling the forged aluminum A-arms after hardcoating was tough: but worth it in the way they look. BTW: Since building the Mule, I have used mostly Detroit Speed front subframes/suspensions as they work great and come painted/preassembled (lots less work and still good looking!).

This subframe used C4 Corvette front suspension components: including the 'Vette uprights with the wheel speed sensors. My intention with the Mule was to create an Anti-lock Braking System (ABS), but I never completed it (my latest car, the Red Devil, is the first '69 Camaro ever to have ABS on it—which I developed).

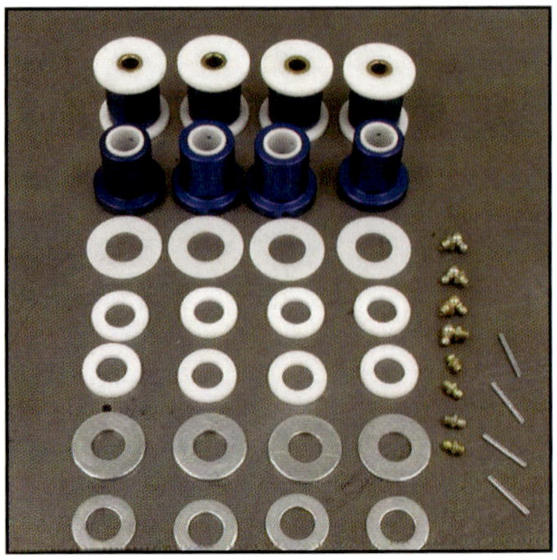

Luckily for me, Global West has a Delrin and aluminum A-arm bushing kit to replace the stock rubber bushings. These bushings are drilled for grease fittings, which allowed me to keep the 'squeaking' aspect of these bushings to a minimum.

Top 3 Final Assembly Mistakes

1. Forcing Parts Together

This seems so simple: if components are 'tight' going together, don't force them. Stop, figure out why there is a tight fit and fix the problem. Unless parts are intended to be united in a press-fit situation, they should go together by hand. Period.

2. Gunky Threads

When tightening fasteners down, if they don't thread down cleanly, stop and remove the fastener Chase the threads down with a correctly sized and thread type tap or die (depending on whether the bolt or threaded hole need help). Spray the hole out with solvent, apply a light amount of oil OR thread lock (depending on what you're doing with the fastener) & thread the fastener in again.

3. Improper Fasteners

I am a stickler for using Grade 8 fasteners in critical fastening situations (which for me is practically everywhere on the car). They cost more, but not much, and offer the maximum clamping capability. I also trim any excess off each fastener so they look 'right' and keep the vehicle weight to a minimum.on that issue and move forward.

The A-arm bushings need to be pressed into the A-arms without damaging them—a very stress-filled, difficult process that requires immense patience to end up with properly installed bushings and brand-new looking A-arms.

It takes substantial force to push the bushings into place, which is why the press is used. I always use large plastic blocks on both sides of the press process to minimize any damage to the A-arm surface.

To press the main shaft bushings in without crushing the A-arm, I build two 'spreader' bars out of angle iron (one is positioned on the A-arm - arrow), then wrap duct tape over the ends to minimize the chances of it marring the surface of the A-arm. The two plastic shims are used to push against in the press.

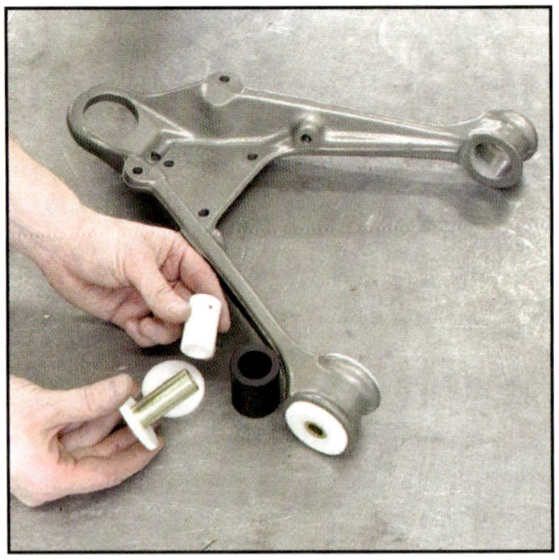

This is how most Delrin bushing systems work: a steel tube insert is nested inside a piece, or pieces, of Delrin plastic, or a similar material. This replaces the factory rubber bushing that has more compliance (and are used because they reduce road noise/harshness transmitted into a car).

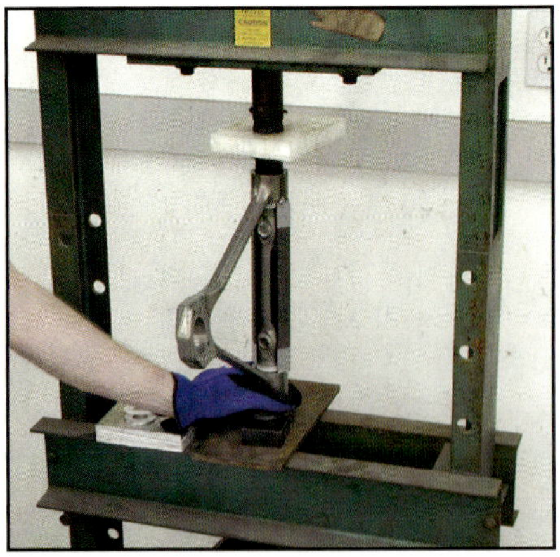

This is serious stuff here. I have one of the Mule's upper A-arms in the press with the two spreaders in place and am pushing on the bushings with the plastic shims. I am always happy when this is completed and I can move on to the boring bolt-on stuff. Use caution when doing this...

The ball joints on most A-arms are pressed in to position…and usually require some type of special shim to install. In this case, I cut away/ground on a piece of 2 x 2 inch square tube so it sits flat on the A-arm (and clears the undulated areas). Then I wrapped some duct tape on the pressure face(s) to minimize the chances of marring the show surface.

The next step is to install the A-arms on the vehicle. As an FYI, I take the paint thickness into account when fitting the suspension in the brackets and how everything interacts. This might sound extreme, but that way, it all goes together with no grinding/sanding/filing.

I used double adjustable QA1 coilover shocks on the Mule. Notice I am using Grade 8 fasteners on the car and the bolts are cut to show 1.5 to 2.5 threads when fully tightened. I use three drops of blue Loctite (pn 262) on the threads of all suspension fasteners.

The Speedway Engineering swaybar assembly was probably the easiest element to final install…after I cut the frame to tuck the arms in, built my own mounts and horizontal retaining clamps and had everything nickel and hardcoat anodize plated so it looks cool. No problem, right?

I prepped the bushings by wiping them up with high pressure lube before sliding the splined bar in place (this minimizes binding issues). Notice the zerk fitting on the bushing mount: these bars need to be lubed occasionally to avoid binding them up.

The Mule was the first tri-year (67/68/69) Camaro where I didn't run a recirculating ball front steering system—and I can tell you it was a revelation. I haven't run one since! This rack took some work to get to fit in the car, but was worth it in steering input.

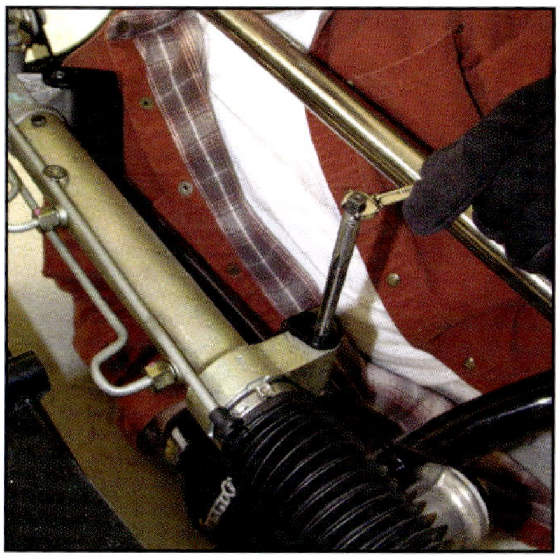

The body on this steering rack had threaded holes in it, but the threads were anything but 'right'. I spent a lot of time cleaning the threads and then sprayed about a half can of brake clean into the hole with a rag over it to get all the metal chips and gook out of the blind hole. I let the brake clean evaporate, then tightened the fasteners down.

I spend a lot of time mocking up ride heights, suspension movement and making sure the suspension is ready for lots of road/track use. Here, I'm checking the clearance of the steering rack/Baer rod ends to the Speedway Engineering swaybar assembly during full compression of the suspension—which will happen during hard braking!

I am final mocking up the Vette bearing carriers on the uprights (thus the odd-length bolts holding it in place). I mark the desired length with a paint-pen on the threads, then cut/grind them to length before reinstalling with Loctite on the threads. Notice the C5 Vette swaybar links connecting the swaybar arm and A-arm: NICE!

Rear Suspension Final Build

It handles the fun stuff: big power and tire smoke, so now make it work

Having built a lot of high horsepower Pro Touring cars and driven most of them ridiculously fast, you'll need to believe me when I say this: the rear suspension needs to be ready to handle hell and back—because that is what it's going to get! Oh yeah, we all say we'll be good to it, tell it everything is going to be OK, but let's be honest. After a few miles of being nice, it'll be burnouts, sliding, hard cornering and drag launches from here to eternity. Sounds like fun, huh?

So with that in mind, getting the rear suspension/rearend installed properly and looking good is definitely an important step—but it shouldn't be hard to do. Just to be clear: you should have already done the hard work, which includes getting the suspension mounts in the car, setting the suspension link lengths by cutting/adjusting them close to the proper length (to position the axle appropriately in the wheelwell and 'square' in the car) and final assembling the differential/axle assembly. If you don't have these things done and decide to have the parts powdercoated/painted anyway, I would expect you'll need to do some grinding/cutting/welding to make everything fit (which is a real bummer, as you just spent all that time/$$$ to get it all looking nice and now you have to mess it all up because you didn't do the proper work up front).

So back to my original thought: if you've done all the prepwork that is required, the final assembly should be a non-event. That is not to say it will be a no-brainer, because since the rear suspension is going to be beat down constantly, you need to give it every chance possible to provide reliable, consistent service.

The Details

By the time I get to this point in a buildup, I have had the rear suspension on and off the car at least 3 or 4 times (usually more than this) and am very confident it is almost ready for primetime. Even with that said, I know there will be surprises. Usually, these revolve around threads that have become clogged with powdercoating/paint/plating or clearances between parts becoming too tight due to powdercoating/paint/plating thicknesses not accounted for in the rough buildup.

The one-off rear suspension I built for the Mule looked really good during its final install. I used quality parts like the Currie aluminum center section 9 inch rearend, Hotchkis swaybars, Reider Racing Gears center section, QA1 coil over shocks and Coleman bushings to make it all 'work'.

To avoid the clogged threads issue, I work with companies that use robust threaded hole plugs when working on my parts (there are plenty of powdercoaters/painters/plating shops that don't do this detail work, so it's your job to ask if they do). On the thickness issue, I usually allow 0.030 inches of thickness for powdercoating, 0.010 inches for paint/primer/etc and 0.003 inches of additional thickness for plating (like chrome/nickel/etc). These are rough numbers, but have served me well over the years, so I use them.

Travel Check

As I did with the front suspension, I like to run the suspension through one last check of 'full travel' to make sure there is no excessive binding in the suspension. I also check that the brake hoses and parking brake cables are all long enough to avoid binding and/or being ripped apart (really a bad deal). I do this by assembling the entire suspension, but with no coil springs in the coilovers. This way, I can cycle the suspension up/down and then twist it up in the car (like the car is leaning through a corner). I like to have at least 1 inch of 'wiggle' left in the hose/cables when the suspension is at full 'droop' (like when you are jacking up the car to change wheels/tires) and a minimum of 1 inch clearance from any component by the hoses/cables.

I run what many would consider to be substantial wheel/tire clearances—like ½ inch to any 'hard touch' surface. I do this because I run my cars so hard I know the tire 'squirm' (where the tire is being heavily deformed from sideway g-forces) is at least 3/8 inch. If I didn't protect for this, I would have tires rubbing the body/frame/suspension parts at very inappropriate times (like the 100 mph sweeping corners at Laguna Seca or Virginia International Raceway!). You can decide whether to protect for this or not: some people are confident they will never see their cars run that hard.

For more on getting a rear suspension together and making it work, check out the following sidebars and photos/captions.

Top 3 Ways to Build Your Rearend Right

1. More Strong

I always build my drivetrains to handle at least 30% more power than I plan on making with the engine. This is why I still use Currie 9 inch solid axle rearends in my cars. They are somewhat bulky, but set up with 31 spline axles and Ford Big Bearings, they can handle 700+ hp at the tire. I don't know of many independent rear suspension (IRS) systems that will handle big, big hp with big, big tires like that....

2. Light is Right

Even with my rule about extra power capability, I am always working to minimize the weight of my vehicles, but especially the suspensions. Keeping weight off usually falls to shortening bolts to show only 1.5 threads out of the nut, sculpting brackets around mounting holes and thinking ahead to minimize size and weight.

3. Good Geometry

Whether you have a solid axle or independent rear suspension, you need a suspension with good 'geometry'—which means the suspension does the best job possible keeping the tires 'flat' on the ground (which provides the most traction) and also leverages the mass of the vehicle for maximum traction (with things like 'brake anti-dive', 'acceleration squat' and 'axle pinion angle wrapup'—which are concepts that are too big to define here...you just need to know they give the tires maximum grip when needed.

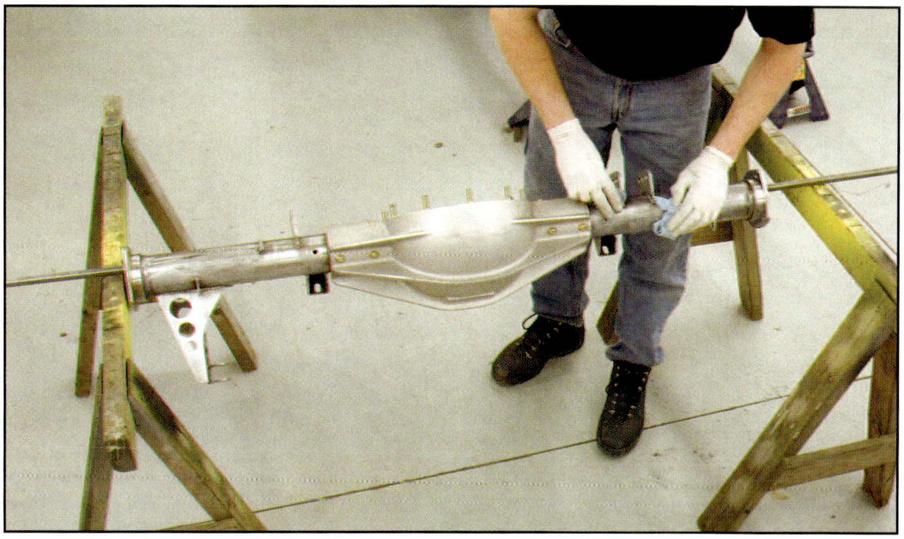

One of the final steps I needed to do on the rear suspension was to have the axle tubes prepped and painted.

Getting the metal completely bare on the axle tubes is accomplished by scrubbing the metal with Scotchbrite pads soaked with a MetalPrep/water solution. This is serious active solvent, though. Make sure you do this process in a well-ventilated area.

I believe in taping off the areas you don't want paint to adhere to. This is an especially good idea when spraying on the ultra-aggressive PPG DP90 primer/sealer.

The DP90LF used here was mixed with PPG DT 870 reducer and DP402LF catalyzer in a spray gun in preparation for applying. If you don't know how to do this, I suggest you pay someone to do this work...which I did here.

Yeah, I know, this isn't a paint booth. I was in the final thrash to get the Mule completed, so I worked with what I had. The shop we painted in had any dust blown out really good the night before and the paint was applied first thing in the morning when the air was calm.

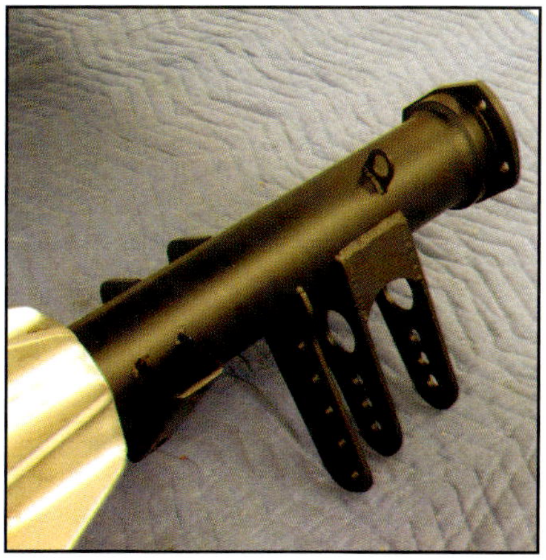

The axle tubes looked good considering where the paintwork was done. Project Buildup Note: I always have to make compromises like this during a buildup and they usually work out: this is a good piece of reality. Sometimes you'll be in situations like this—my advice is to do the best with what you have and keep moving forward.

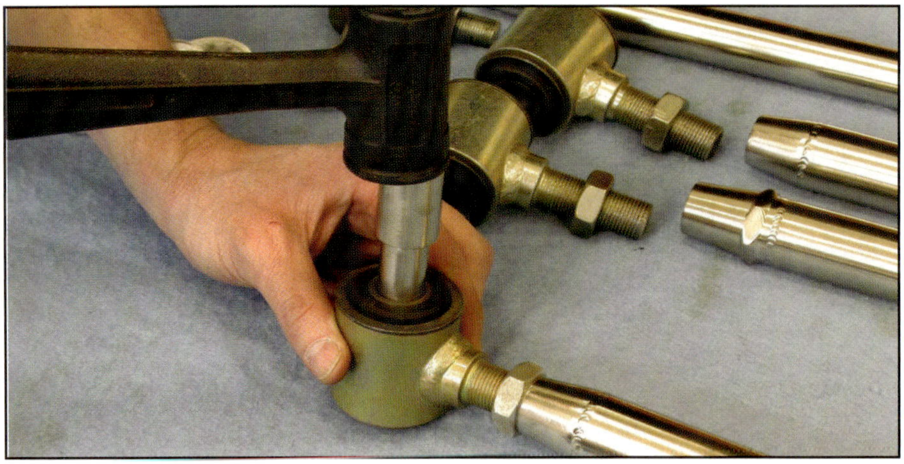

I install all the bushings and metal inserts at this time. These are coated with high-pressure lube before being assembled to minimize the chance of 'binding' and also to limit any squeaking.

I like to start installing the rear suspension by loosely bolting the suspension arms to the car. All of the lengths were set in the last mockup and documented. I preset those lengths before loading them into the car.

Once all the suspension arms were in place, I loaded the Currie rearend up under the car and set it on jackstands. Then I started bolting up the suspension links (top first, then bottom, then panhard bar) before installing the coilover springs.

Once I got the suspension all bolted in, I begin to set the suspension to 'ride height'. Again, these parameters were established during the mockup phase and I just replicated these settings at this phase of the buildup.

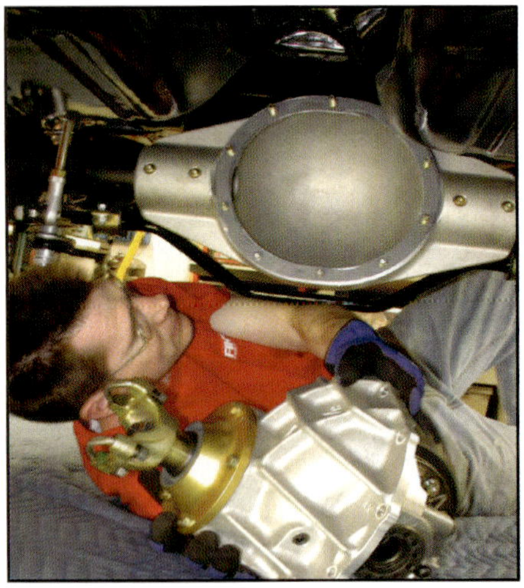

The last step for installing the rear suspension is loading the differential into the Currie 9 inch and sliding the axles into the housing. This is why I run the 9 inch rearend—it is super easy to assemble and if I decide to make a gear change, it can be done very easily.

I then bolted on all the Fikse wheels/BFG tires and lowered the vehicle off the jackstands to check my ride height settings.

So, here is the Mule as it approached its completion. Looks pretty good, doesn't it? (Hard to believe this car was completed back in 2001). I decided to raise the rear of the vehicle a little after getting this look; to account for the weight of the fuel in the tank.

This is the advantage of using coilover shock absorbers: I just jacked up the rear, set the car on jackstands, lifted the rearend up slightly with the floorjack and rotated the spring mount nuts up 4 turns to raise the car ½ inch. Leaf springs are waaaaaayyyy harder to adjust ride height on!

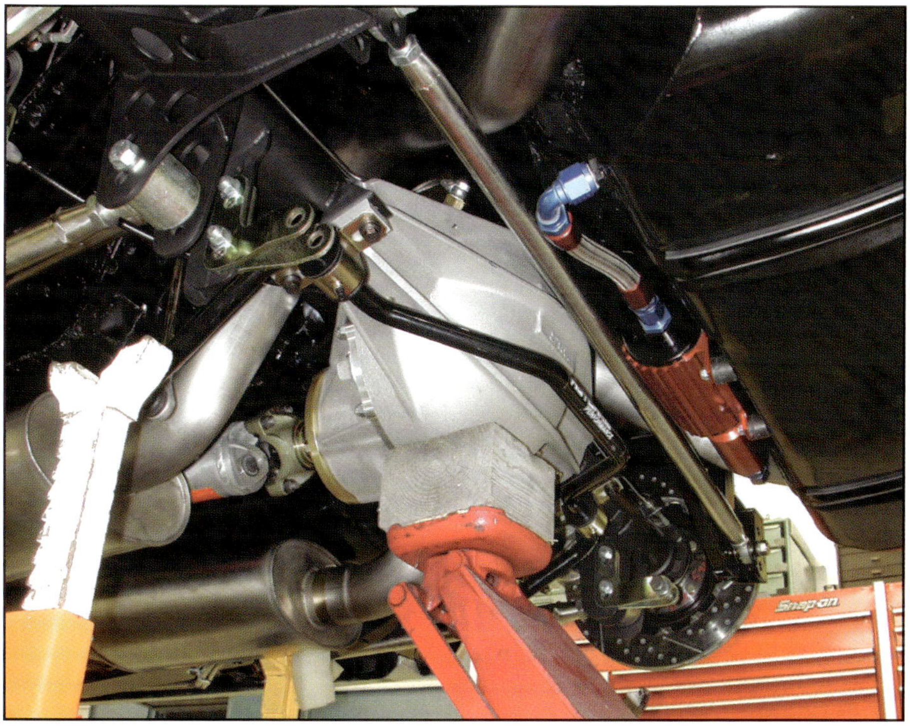

Here is what the rear suspension looked like when it was all completed. Notice how the Stainless Works exhaust fits cleanly around the rear suspension and the Aeromotive fuel system is tucked just in front of the panhard bar.

I am very proud of multiple aspects of this suspension: it ended up working exceptionally well on the street and track, looks great and pushed the aftermarket to sell rear suspensions for Pro Touring cars that actually 'work'. I also hope it encourages people to experiment more with their builds.

Brake System Final Install

Setting it up to STOP

Since I build musclecars with big horsepower and grippy tires, having a brake system that has enough capability to stop the car multiple times from big mph numbers is key to having safe fun. To do that, I run big disc brake and caliper systems that actually work. The Mule had a full Baer Brake combo which worked quite well for the abuse I heaped on it. No matter what manufacturer's parts you choose to run (I have run brake systems from Wilwood, Baer, Brembo and others), it is critical to get the components sized correctly in the system, then assembled correctly and finally, get all the air bled out of the system.

While I like a brake system that looks good (especially what you're going to see through the wheel openings), the brakes need to 'work' exceptionally well on my Pro Touring cars before anything else.

Top 3 Brake Install Mistakes

1. Twisted/Short Hoses

The flexible hose that connects the frame mounted-tubing to the caliper at each corner of the vehicle needs to have enough length in it so the suspension can move through its entire travel while keeping some 'slack' in the hose. The hose also needs to not rub on the wheel/tire, suspension or anything else—which can be a challenge when you have extra material.

2. Burnish 'Em

When you first get your vehicle up and running, I suggest the first step you do is 'burnish' the brake pads/rotors. To do this, I perform 8 stops from about 35 mph using mild pressure (these are NOT emergency stops). Then, I take the vehicle up to 45 mph and do a 'hard' stop. I repeat this hard stop segment 2 more times. I then drive around for about 20 minutes without touching the brakes to allow them to cool down. Doing this process transfers some brake pad material to the rotors, which is how the brakes provide maximum braking capability.

3. Keep it Clean

Installing rotors that are built with a zinc-coating will make it easy to keep the rotors looking good. While I don't drive my cars in the snow and salt of winter (I live in the Midwest), if I used uncoated rotors they would get rusty just from washing my car or driving it in the rain. Since that is unacceptable, I always get rotors that have the protective zinc coating...seems like an easy choice to keep everything looking good.

Mule Brake System Rundown

1. *Baer Brakes Disc brake system*
2. *Inline Tube hardline (stainless steel, 3/16 inch OD)*
3. *Earl's braided steel flexline (from framemount to caliper)*
4. *Earls check valve bleeders (to be able to bleed system with one person)*

Top 3 Brakeline Survival Tips

1. *Locate hoses and hardline up where off-track excursions won't damage them*
2. *Position hoses/hardline away from heat sources (exhaust, cooling hoses)*
3. *Make sure hoses/hardline are not rubbing on anything during vehicle operation*

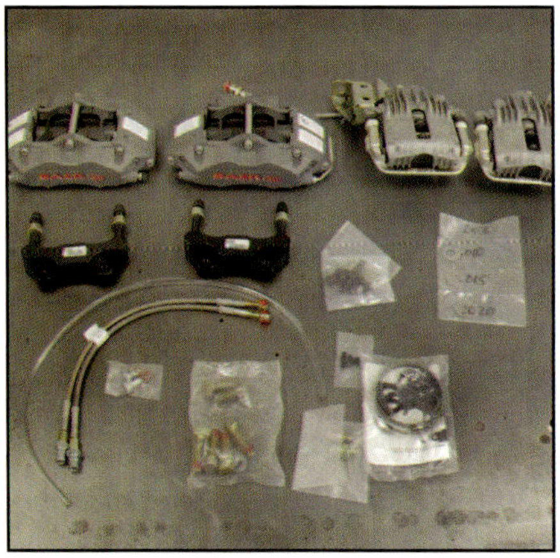

The Mule was equipped with a Baer disc brake system that came with everything you see here plus a set of zinc-dipped rotors and 'hats' that allow the race-style rotors to mate to the 'Vette wheel hubs. The zinc-dip on the rotors minimizes surface rust from appearing on them.

The first step to installing most aftermarket brake systems is to bolt on the bracket that will adapt the performance-style brake caliper onto your spindle upright. You will need to measure that this bracket is 'square' to the face of the rotor (it's not that hard—I describe how in next caption) so the entire brake pad is coming into contact with the rotor.

Once I have installed the rotor on the hub and the caliper on the bracket, I measure the 'gap' between the caliper and rotor at the top and bottom of the caliper. This distance should be the same. If it isn't and you don't fix this, the brake pads wear oddly, the system will not provide maximum braking capability and there is a good chance the brake pads will 'drag' .

Getting the calipers plumbed for brake fluid usually requires some custom fittings and hose, like this brass -6 AN flared end fitting and 3/8 inch flared male opening adapter. This is used to mate up the brake hose to the hardline and position the hose end in a framemount.

I purchase framemount brackets from Coleman Racing in Coleman, Michigan, and weld them to the frame before I have it powdercoated. Coleman offers just an insane array of brackets and parts that come in handy when building a Pro Touring car.

The brake hose is connected to the caliper with a Baer-supplied 'banjo' bolt (these are centerdrilled in the threaded portion of the bolt and side drilled just under the head of the bolt). A brass washer is placed on each side of the hose fitting to help seal it to the bolt/caliper.

Here is the banjo bolt/brass washers/hose fitting assembly bolted to the caliper. Notice how the hose is 'clocked'—I spend a lot of time trying different hose arrangements until I find one where the hose stays 'slack' throughout the entire suspension travel and doesn't come in contact with any of the moving parts (not easy to do!).

I modified a Detroit Speed & Engineering dual master cylinder to operate without a vacuum assisted booster on the Mule because the rollcage downbar was in the way. Because of this, I decided to add this safety loop (pointing to in photo) for the brake pedal pushrod.

The rear brake assembly for the Baer brakes uses a large bracket that bolts to the Currie rearend to hold the caliper over the rotor. I perform the same caliper location measuring/tuning and checking of the flexline length/position as I do on the front.

Building Brake System Hardlines

While it takes a long time and requires a lot of patience and skill, I build my own hardlines for the brake system. This starts with me buying about 30 feet of 3/16 inch stainless steel tubing, along with about 30 male and female 3/16 inch brass and stainless steel fittings and flexhose for sheathing the tubing.

The only way I know to do this job is to start at one end of where you want to plumb to and work to the other end: in this case, we are working from the right front corner to the balance bar/master cylinder. Notice the packing blanket being used to make laying on the floor more comfortable!

I think the best way to bend hardline to fit a car is to use an angle-guide to determine the desired angle, mark the tube with a 't' shape to indicate the bend centerline and then bend the tube away from the fitment area. I reinstall the tube under the vehicle to check the bend and prepare for the next bend.

I use a cutoff wheel to cut my hardline. It provides a quick, clean cut that requires little physical input by me: don't laugh, bending up hardline is an exhausting job that requires sharp mental skills. When I get tired, I make mistakes: so I work to stay fresh.

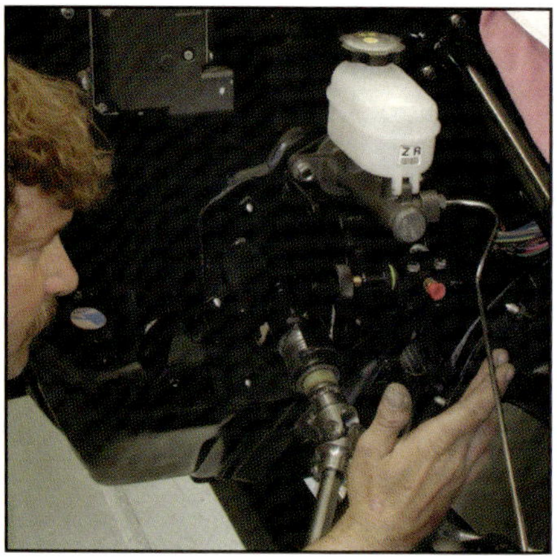

I get very focused as a hardline approaches its connection point, as I know there is a lot of work already into the job—make a mistake now and its back to square one. I have a fitting installed in the master cylinder to be able to mock up the tubing (make sure to leave enough length on the tube to apply the flare for the fitting!).

Sometimes, you'll need to 'tweak' the tube a little to tune up a bend or its angle. I do this a lot and have found you can put in about 5 or so degrees by hand without making the tubing look 'wavy'. Add any more than this and the tube will look like you didn't know what you were doing…

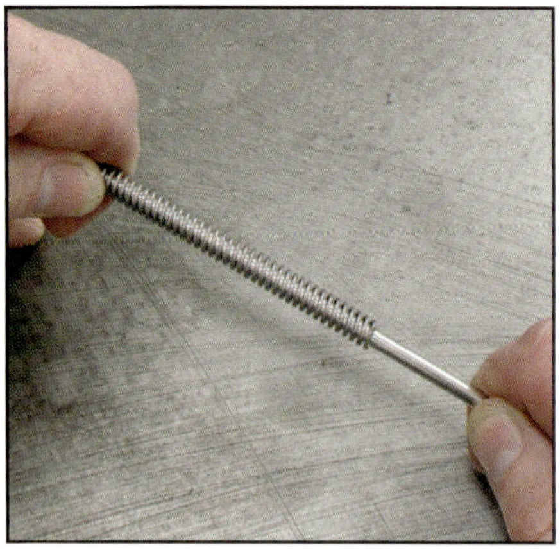

I like to use the factory coiled sheathing around my hardlines—my opinion is if the factory used it, there must be some really good reasons for it. I buy if from Inline Tube, and slide it onto a tube just before I add the second flare.

I built the Mule using this manual flaring tool, but have since bought a hydraulically powered one (I got tired of the cranking…). The key to flaring is making sure you have the nut on the hose, the 'right' amount of tube protruding from the clamp (I match the thick end of the form tool) and that the form tool compresses straight down on the tube to create the flare.

Here is what the flare should look like after you have cranked the form tool down until it bottoms on the tubing clamp. Notice the nut just poking out from behind the clamp—make sure you put this on before doing the flare!

This is what a quality flare looks like. Both stainless steel and mild steel tubing should be flared with a 45° flare tool.

The final step to installing the hardline is determining mounting locations for the small tabs that hold it in place on the frame. I like to use these little billet aluminum tabs held on the frame with stainless steel button head Allen bolts.

Once the location is determined, I drill a hole in the frame section.

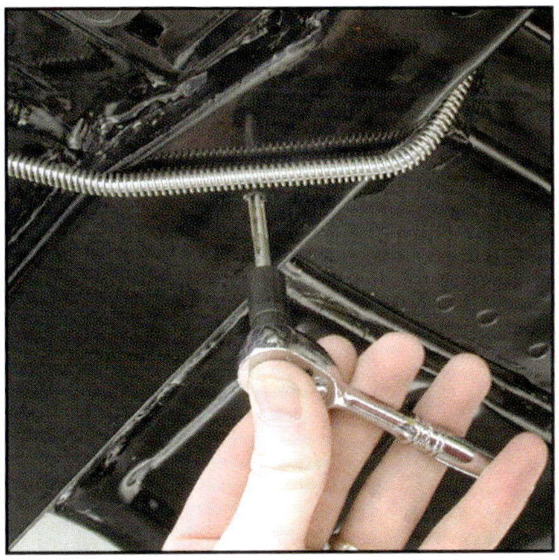

Then I run a threaded tap up the hole to allow the bolt to be installed. I apply a copious amount of thread lock to the bolt and install the bracket/bolt into the frame—making sure to use a minimal amount of torque on the bolt (as it is easy to strip the threads out).

When installing the flared nuts into the flexhose fittings, I like to use thread sealer, like that available from ARP, on the backside of the flare/frontside of the fitting. This is just extra insurance to get a complete seal at this critical joint.

Front Engine Accessory Drive (FEAD) Install

The easy way

This was one of the nice surprises I experienced on the Mule buildup—and I recommend everyone do something like this to avoid hassles down the road. This is the installation of Vintage Air's Front Runner accessory drive system—it actually installs as advertised and works like you'd expect (you can install it and forget about it).

And, let's face it, this system is a huge departure from what most of us have known in our lives: the cobbled together combinations of stamped metal brackets, V-drive pulleys, spacers, washers and other violations of engineering. And usually, all of it snarled together in a last minute thrash of insanity.

The System

Well, kiss those frustrating memories goodbye with the Front Runner system. This single-belt serpentine drive system was designed as a computer model first and tested in the computer to make sure everything worked as expected—long before a single machining chip was whittled.

Vintage Air chose to base the entire system on a single, massive main structure that is CNC-machined out of a solid chunk of billet 6061 aluminum—and weighs only 3.5 pounds! There are two smaller brackets also machined out of 6061 aluminum billet, as are the billet pulleys—all developed in-house by Vintage Air's engineering team.

The fastening duties are handled by ARP studs, bolts, nuts, washers and other hardware. In the end, this system looks completely 'right' in any performance situation.

Vintage Air offers the system with the A/C compressor, alternator and, if you need it, the power steering pump. This way, they can maximize the chances everything will bolt on. They even recommend using a Stewart water pump and their own special 'bypass' thermostat (there are instructions showing how to drill three small holes in an existing thermostat to make it into a 'bypass' one if you already have a thermostat).

I liked how easy the system was to install and after beating the hell out of the Mule for thousands of miles and not having a single issue with the Front Runner system, I can say it does all one would expect—spin the accessories!

If you are running a Gen 1 small block V8, you want to run the Vintage Air Front Runner system—it installs easy, looks great and works.

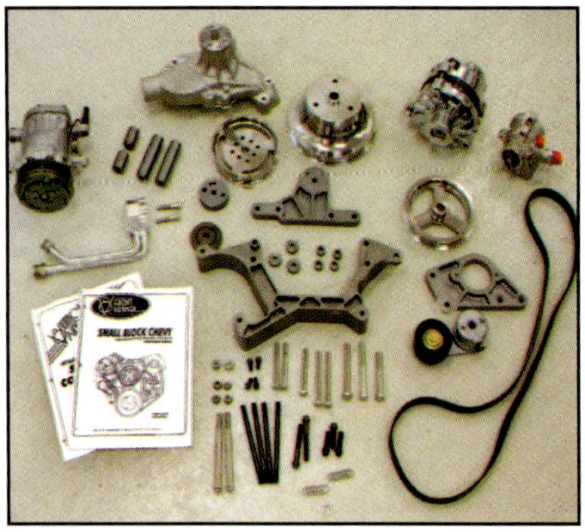

There have been many accessory drives built by the aftermarket, but this system reset the bar when it came out. Just look at that billet aluminum main mounting bracket! This is what excellent engineering looks like.

Again bucking the norm, the directions for the Vintage Air Front Runner system are excellent. The illustrations, text and sidebars make it easy to follow the steps required to install this system. I do like to 'mock-up' everything to check it before installation, but this kit went together so well, that was really not needed.

OE or Aftermarket?

Most of the Pro Touring cars I have built over the years have had a serpentine belt FEAD that was custom built using production brackets and accessories (alternator, A/C compressor, power steering pump, water pump and belt tensioner). I would go to the junkyard and come home with piles of brackets, pumps and pulleys and then go to work modifying them until they worked together as a system.

These systems were incredibly durable and looked good under the hood, too. But they took absolutely forever to build.

I have always been looking for an alternative offering in the aftermarket, but never could find a system that was up to my standards. I would tell my friend Jack Chisenhall, of Vintage Air, how this was a big problem and that someone needed to offer a system.

Well, Jack must have heard me because he had his team build the excellent (in both how it installs, works and looks) FEAD system you see being installed in this section and it has become the required system for anyone building a Gen 1 small block-powered car (which is still a great engine that continues to become more and more affordable by the day). This system worked like a charm through immense abuse and never whimpered once—so thankfully, my days of building FEADs is over!

This is one of my secret assembly tricks: I like to work on 3x4 foot, 2 inch thick foam pads when assembling systems. This allows me to get off my feet, minimize the chances of marring the surface of what I'm assembling and makes it easy to access everything I have laid out (on the floor).

Any good A/C compressor will be shipped with a block-off plate—remove this as the first step to building the Front Runner system (this will not be reused).

The Front Runner hardline distribution block was then bolted on with Vintage Air's supplied 40 mm long, 8 mm diameter bolts. These were torqued to 10 ft-lbs per the instructions.

There are two 1-inch outside diameter (OD), 5/16 inch inside diameter (ID), 3/8 inch thick aluminum shims included in the Front Runner kit. They are used to fill the space between the mounting tabs on the Sanden A/C compressor and the Front Runner main bracket system. Another spacer that is 0.280 inch thick with a 3/8 inch ID and 1 inch OD is used between the alternator and the main bracket. Two 8 mm diameter , 90 mm long bolts are threaded from the rear of the main bracket to hold the A/C unit in place.

This is refreshing: the 8 mm diameter, 75 mm long bolt that threads into the pivot mount of the alternator is JUST long enough to engage all the threads, but doesn't hang out of the alternator body needlessly.

This is the backside of the Front Runner main bracket. Notice how the A/C compressor and alternator nest perfectly into the bracket—this is the result of excellent engineering work!

The main bracket for the Front Runner system is anchored on four ARP 3/8 inch diameter, 5 inch long studs that run through the water pump mounting holes. I suggest you apply a bead of thread sealer or silicone to the threads before installing these in the block (to minimize the chances of coolant leaking through them).

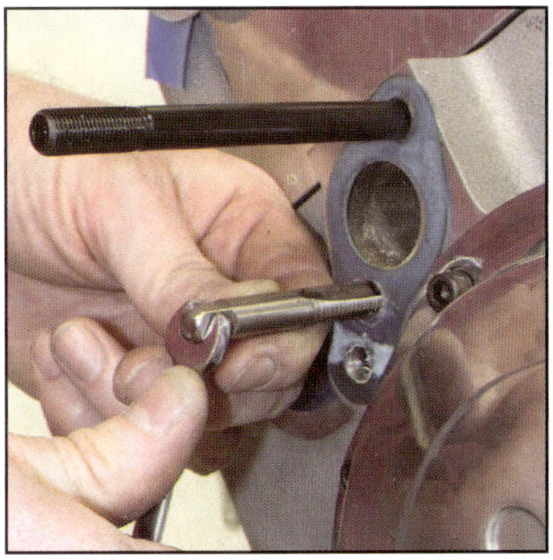

There is a good chance you'll have to clean the threads on the water pump mounting holes to get the studs to fully seat in the holes. Remember, 1 turn in, ½ turn out, repeat—this cleans the tap threads so it can do a great job cleaning up the threads in the hole!

As usual, I modified something on this deal—my billet aluminum Summit Racing timing chain cover is thicker than the stock stamped one—to clear it I swapped the stock hex bolts on the Stewart water pump, for 5/16 diameter, ½ inch long button head Allen bolts (right).

After installing the Stewart reverse flow water pump (after applying a bead of sealer on both sides of the water pump gaskets), I slid the four 0.395 inch thick aluminum spacers onto the main mounting studs.

The next step is to slide the power steering bracket onto one of the main studs on the driver's side of the engine and thread the two 12-point, 3/8 inch diameter, 2 inch long, coarse thread ARP fasteners through two 1.5 long, 1.0 inch OD spacers.

The main bracket, with the A/C compressor and alternator bolted up, was then slid onto the main mounting studs. Notice how the A/C compressor hardlines run behind the system: slick, huh? The alternator was bolted to the engine & power steering bracket

The idler bracket was then bolted on using 4 inch long ARP 3/8 inch diameter bolts and 3.675 inch long Vintage Air-supplied spacers (these nest into the backside of the main bracket). The billet parts are 'hard coat' anodized, which makes them look very nice.

Once the idler bracket was installed, it became obvious that I had installed the main studs too deep in the water pump holes to get a washer/nut combo on them…so I backed the studs out just enough.

The power steering pump needs to have the Front Runner pulley installed on it at a specified depth. I used a pulley installer (available at most auto parts stores) to do this.

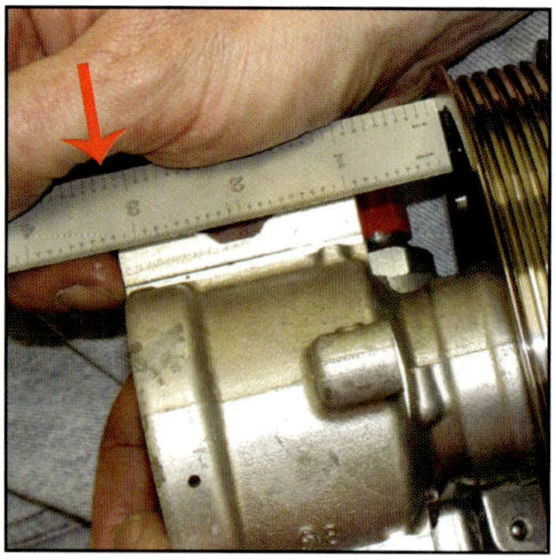

The backside edge of the pulley needs to be located 3.2 inches away from the rear edge of the pump body (see arrow) to line up this pulley with all the other pulleys.

The power steering pump was then positioned over the holes in the Front Runner bracket and the fasteners hand threaded in from the backside. They were then torqued to 30 in-lbs.

After installing the crank pulley with three 1 inch long ARP 3/8 inch bolts, I substituted an ARP crank bolt (in place of the nice Grade 8 bolt that comes with the Front Runner kit) as it has a cool hex-shaped hole in it that I like to use to turn over the engine.

I then slide on the water pump pulley with the deeper section of the pulley going over the water pump and bolted it on with four 5/16 inch diameter, ½ inch long ARP bolts.

To install the idler pulley, a provided dowel pin needs to be carefully installed in the bracket—I suggest using a brass hammer (shown) to tap it into place. This dowel allows the idler to create a torsional force against the belt to hold it tight on all the pulleys.

There are two tensioning holes (arrow), one for use with a power steering pump, the other for without the power steering pump.

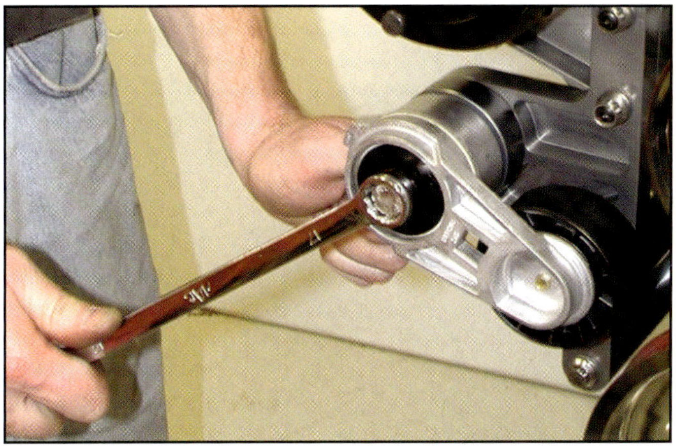

The idler installed by assembling the spacer on first, then the idler as shown before bolting it all together by tightening a 17 mm bolt/nut combo down. This is how the idler should look on a power steering-equipped system.

This is a good way to check your work. All of the pulleys on the Front Runner system should line up within 1/32 inch of each other with everything installed.

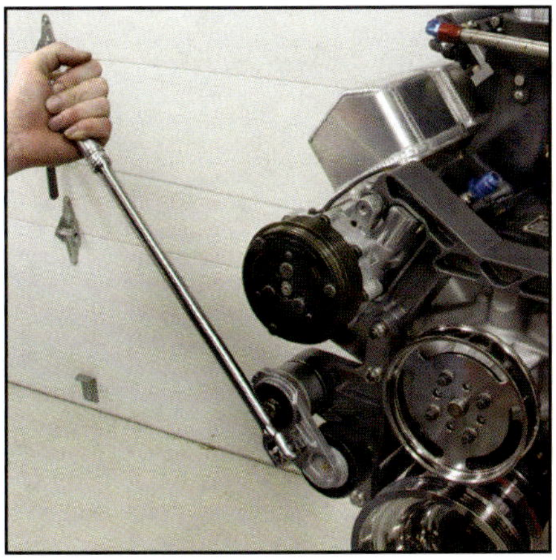

To begin the install of the belt, I like to get a large 1/2 inch breaker bar located on the belt tensioner to see which way I need to pivot it to 'relax' the tensioner to get the belt on (in this case it is pushing the breaker bar up, or clockwise, to relax it).

Following the Vintage Air instructions, I then threaded the belt around onto the various pulleys until the one are that needed to be installed was just in front of the water pump.

I then relaxed the belt tensioner and slid the belt under the water pump pulley to final install the system.

I then went back over all the fasteners to make sure everything was tight before calling the install complete!

As I said, this is the nicest front accessory drive I have ever built on a car…and even better, it was one of the quickest I have ever put together, too.

One of the nice features of this system is the understated, but classy, appearance. This is apparent in the small engraved logo on the top of the main bracket. Looks good, but isn't so big it looks crass. Nice.

HVAC Details Finalized

Keeping it cool while keeping cool

There are so many details to building a hard charging Pro Touring car that demand my attention, you'd think having a Heating Ventilation and Air Conditioning (HVAC) system that blows cold in the summer and hot on cool autumn days is not high on the list. But you would be wrong. I absolutely demand A/C in the summer—especially when I'm at track days. Sure, I shut off the A/C when I'm on track, but when I'm sitting in the paddock, I've got it on High Cold: and enjoying the luxury.

So how do I get the HVAC to work like that on an ultimate Pro Touring application? Well, there are many details. But the main elements to success are starting with a quality aftermarket A/C system (Vintage Air), then giving the vehicle enough cooling capacity to easily handle max A/C cooling with the engine spitting off maximum heat to the radiator. Sounds easy, but really, it's about as hard as anything I do on these cars.

I have installed air conditioning systems in all my Pro Touring cars since 1993. You know why? Because I can! I use Vintage Air products because they do a great job cooling the vehicle. If you want to have maximum cooling capabilities, you need to make sure the radiator on your vehicle is big enough to cool the engine while the A/C is keeping you cool.

The A/C system

I used the Vintage Air Gen II Compac system on the Mule—in fact, this system is the same one I have used on almost all of the builds I have done since 1994 . I like it because it is compact enough to fit under the stock dashboard, has modern controls and has enough capability to cool the interior on even the hottest days. I used to have to build a bracket for my cars, but now Detroit Speed & Engineering (DSE) offers a bracket (pn 001011) that you can purchase to make installation easy under the dash of a 67-69 Camaro.

The CRFM

In the production automotive business, they refer to the 'Condensor, Radiator, Fan Module' or CRFM, when talking about the cooling capability of a vehicle. The reason they do that is because these components need to work together as a team to provide heat exchanging capability to maintain engine temperature and allow the A/C to blow cold on smoking hot days.

At this point in the Mule build, I knew I was going to increase the engine's horsepower output, but I wasn't sure to what level. Because of this, I sized the Griffin radiator to what I thought was a 'safe' number of 800 hp (we ended up making over 1000 hp on the dyno). I am a big fan of big fans, and the Mule had the highest CFM (cubic feet per minute) fan I could find: the Vintage Air Monster Fan.

I trimmed the fan shroud and built some aluminum brackets so the shroud would fit tightly to the backside of the radiator.

While you would think the '69 Camaro front end would have plenty of cooling air running through it, I have always had to work hard to provide enough cooling capability for my Pro Touring cars. Because of this, I helped to develop the TechAFX Bottom Feed Spoiler System that I introduced on the Jackass Camaro. This system directs air through the radiator that would normally just go underneath the vehicle—and in testing, I saw a reduction in engine temperature of 7° F. That's a lot for just installing a spoiler!

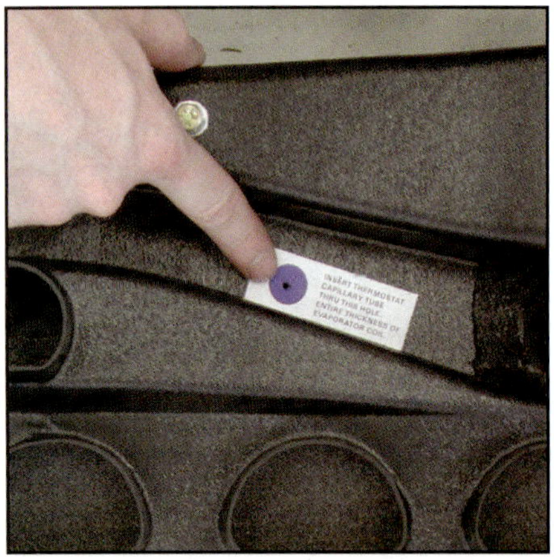

One of the reasons I like the Vintage Air product is that you can set the system to cool the vehicle at a level and it will maintain that—one way it does that is through this hole. This allows the installation of a tube that 'reads' the temperature inside the unit.

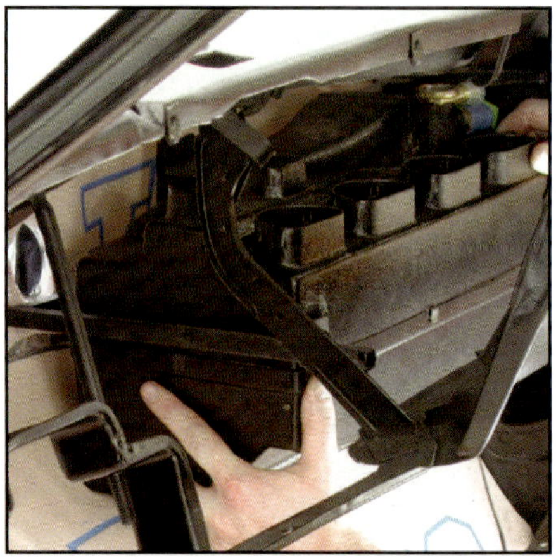

The Vintage Air system is small enough to tuck behind the stock dashboard of the '69 Camaro. This works because we installed a DSE firewall fill plate (pn 010901) to cover the hole left when you delete the stock HVAC system (see page 39 earlier in this book).

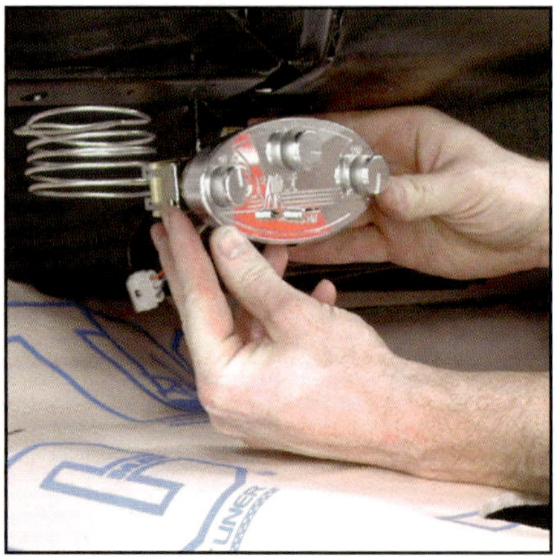

The billet control plate for the Vintage Air system was installed on a small mount I created to fit under the dash on the Mule.

Another cool feature of the Vintage Air system is this electronic two-way valve that is actuated via a switch on the control panel to adjust the flow of hot coolant into the heater core.

Once I had the system installed under the dash, I went to work building the hoses to connect system together. I ordered the hoses and fittings from Vintage Air and pushed a hose on the fitting to start.

I then test fit the fitting/hose on the car just to get an idea of how everything goes together. Once it looks like the fitting/hose will work in the situation, I remove it from the vehicle to be able to crimp the fitting collar onto the hose.

Then I put the fitting collar in the crimping tool I got from Vintage Air.

Once I have fully crimped the collar, I release the tool/fitting. This is what the collar looks like at that point. From here, I take the hose length with one end on it over to the car.

I then apply the Vintage Air-supplied sealer on the O-ring surface. This is critical as leaking A/C hoses are not good for many reasons: the system doesn't work, the R132 leaking into the atmosphere is not good and it costs a lot to recharge the system.

I then run the hose in position up to the other fitting I want it to connect to and mark the hose length. Then I repeat the crimping process again on the other end to complete the hose and move on to the next hose assembly until it is all completed.

So here is the other aspect you need to have if you want a new-car capability A/C system: excellent heat exchanger capacity. To do this on the Mule (especially after I decided to go to twin-turbo power), I upgraded to an aluminum Griffin crossflow radiator.

This Griffin radiator is equipped with a power steering cooler in one side tank (accessed with -6 AN fittings) and an oil cooler in the other (-12 fittings).

I then mocked up the Vintage Air Monster fan and shroud package onto the back of the radiator. This showed a section of the fan shroud needed to be trimmed away. A quick way to mark the trim line in a situation like this is to place the part on a flat surface and pull a straightedge down the face of the part.

I then use a rotary saw to cut the edge. A good tip here is to hold the tool against the edge of the table and slide down the face of the table to make an even cut.

Note the gloves: I always use gloves, safety glasses and ear protection when doing this type of work.

I then 'dressed' the edge of the plastic shroud with a fine tooth file—pushing the file towards the inside of the shroud to knock the edge off it.

I built two aluminum brackets to hold the shroud tightly on the back of the radiator. I bolted them to shroud with small button head Allen bolts and nylock bolts.

To mount the shroud on the radiator, I drilled small holes in the top and bottom closeout plates and installed small RivNuts (these install in a similar fashion to rivets and provide a threaded anchor) for the shroud bolts to thread into.

Here is the shroud being bolted to the radiator with the small Allen bolts. I use thread lock on these so I don't need to tighten them excessively (which RivNuts don't handle well—they spin pretty easily if you try to apply excessive torque).

The next step is to get the A/C condenser installed on the front of the vehicle. I build these simple side brackets to mount the condenser and close-out the opening in the radiator core support.

Here is a shot of the condenser installed on the core support (arrow) and me threading an ARP stud into the radiator mount holes—this makes it very easy to get the radiator mounted in place.

Installing a pair of ARP studs in the core support helps a lot as the radiator assembly is heavy and having to lean over the front of the car to load the radiator into the engine bay is difficult. Notice the many pads and towels on the body—this is how to minimize the chances of marring the paint!

As the final step to getting the cooling system together, I built a set of radiator hoses. This step usually starts with mocking up a shape with a piece of wire, then going through the local parts store hose assemblies to find one that looks like it. Then I install it and trim it to length (shown).

Whenever I can, I work to make my life easier when building a Pro Touring car. One way to do that is to use adjustable components, like this CSI thermostat housing that can rotate 360 degrees.

Doing the same process for lower radiator hose requires more crawling around, but it is just as important.

Once the entire cooling and A/C systems are together, I take the vehicle to a local A/C shop to have them pressurize the A/C system with refrigerant.

Braided Hose Buildup

Plumbing it up

Building the braided steel hose assemblies to plumb a car is always a bittersweet moment for me. I know whatever project car I am building is getting close to being on the road, but at that point I just want to be done and driving it. It doesn't help that the work is very challenging mentally and physically, either. So my coping mechanism for getting through this point in the project is to absolutely throw myself at it. I mean, I just pour myself into the buildup to the point I barely do anything else—in fact, I often take vacation from work so I can do the plumbing work all day and night until it is DONE.

This probably sounds insane, but I know how important the plumbing (and wiring, which is coming up…) is to the reliability and capability of the entire vehicle and how complex it is to do right. And, I have tried jumping in and out of the plumbing process—it doesn't work. Doing this job in pieces will only create problems. In fact, if you do this job in pieces, I guarantee you'll forget some aspect of what you were doing and why and take more time figuring out where you were in the process than it would take to just get the job done.

Building braided steel/AN fitting hose assemblies is a requirement when building a top-flight Pro Touring musclecar. This is how I build mine.

Hose, Fittings and Tools

When plumbing a vehicle, I know things go easier if you have a variety of hoses and fittings laying around. To this end, I have acquired fittings over the years that I store in plastic bins I place next to the car during this process. Some of the fittings are 'used up' ones that I just keep around to use as mockup parts. Other fittings are perfectly good for final builds. Having an assortment of fittings only makes this job go easier as you can quickly try multiple angles and sizes to determine what is best for a situation.

This is a good point to clarify: most of the process of creating hoses for a vehicle is cut-and-paste. You literally just start at one end of a hose and build the hose in position to mate up with the other component you need to transfer fluids to. This is a thinking-mans job that can be very frustrating. All I can say to that is every hose you build makes it easier for you to do the next one. Have fun!

AN Fittings/Hose vs Hose/Clamps

When I was young, I built cars with regular black nitrile rubber hoses that slid on hose barbs and were held in place with the common screw-type hose clamps. But, once I got to the point my cars were going really fast (which happened pretty early on), I decided it was prudent to step up to AN fittings and braided steel hose. Yes, the cost is probably 5x that of rubber hose, but if the rubber hose fails prematurely and you destroy the car sliding in your own fluids, how much did you save?

The New Braided Hose

While the Mule was built with the latest braided steel hose that was available in 2001, time has marched on. The new AN hose is still covered in braided steel, but now it has PolyTetraFluoroEthylene (PTFE) hose on the inside of it instead of the old nitrile-based rubber. I made the change to Teflon, or PTFE, because today's fuels are so caustic that they eat the black rubber hoses from the inside out—quickly causing them to harden, crack and leak. And besides the obvious safety issues, let me tell you, nothing is worse than having a bunch of black rubber gunk pile up in your injectors: they don't work so well after that!

The one outfit that has a complete array of black anodized AN fittings and smooth-bore, conductive braided hose is TechAFX. Their products go together like the rubber-style AN fittings and the hose is impervious to all fuels. Their PTFE hose (which most people refer by the trade name 'Teflon') is slightly stiffer than the rubber hose (which means it can't bend in as tight a radius as the rubber hose) but it is lighter than the rubber braided hose and has been tested to over 2200 psi in hose assemblies. TechAFX offers hoses and fittings in -6, -8 and -10 sizes (they are the only co. with -10 PTFE smooth bore hose), so they can handle just about any fuel system need.

Top 3 Plumbing Mistakes

1. Place It Right

Positioning a hose assembly near moving parts or extreme sources of heat is a no-no. My rule is to keep the hose at least 4 inches away from either of these problem-makers.

2. Cut Perpendicular

I like to use a cut-off wheel on a die grinder as the cut is done quickly and the pressurized air is doing the work—so I can focus on making a clean, perpendicular cut in the hose.

3. Keep it Clean

There are many ideas on how to clean the hose out after cutting it like it do: I have always washed the tube out with solvent and sprayed the hose out with dry, pressurized air and installed the hose...and have had zero debris problems over the years.

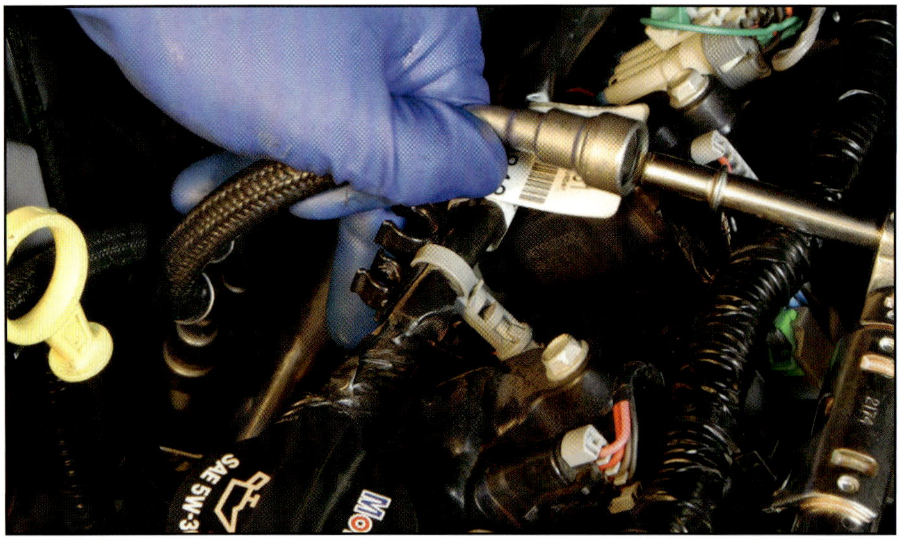

The Future of Hoses

This is just my opinion, but I think the future of aftermarket performance hoses is going to include the type of Quick Disconnect (QD) fittings that have been used on production cars for the last 10 years or so. These QD fittings are so easy to work with and are absolutely bulletproof in their performance, I believe it is just a matter of time before they become available for custom applications.

My latest car, the Red Devil, has somewhat expensive QD-equipped hoses on it that I had custom-built to fit my application, but more and more companies are starting to offer QD-fitting equipped hoses as off-the-shelf items. One of those companies, TechAFX, offers underhood dressup items for the 2010 Camaro and C5 and C6 Corvettes and will be offering more for the Pro Touring crowd in the future.

Fun Fact

Do you know what the 'AN' stands for in AN fittings? Well, it stands for Army – Navy, as these two entities agreed upon standards for all hose fittings back around World War II. Those specifications separated Hose Inside Diameter (ID) dimensions into 1/16 inch increments and represented them with simple numbers that follow a '-': as in a -8 (pronounced 'dash eight') fitting equaling a ½ inch (8 x 1/16 inch = ½ inch) ID hose fitting. AN fittings use a 37 degree flare on the inside of the fitting. So now you know!

AN (-) Sizes	Equivalent Hose ID
-4	1/4 inch
-6	3/8 inch
-8	1/2 inch
-10	5/8 inch
-12	3/4 inch
-16	1 inch

Pressure Testing for Perfection

The most frustrating part of building hose assemblies for me used to be determining whether I had built a 'leak-proof' hose. I say 'used to' as I have found a solution to this problem. I got an Earl's Pressure Test kit (available through Summit Racing, pn D016ERL) and use it on every hose assembly I build.

Here is the process I use: after building a hose, I thread in a blank Test kit plug on one end of the hose, and on the other end I thread in an air nipple-equipped Test kit plug. I then apply 100 psi of air pressure into the hose assembly and leave it for at least 12 hours (usually overnight). I expect the hose assembly to lose no air pressure over this time. If it loses air, I rebuild the hose.

So what leaks? Well, it varies: sometimes it is the mating point of a fitting to the hose, other times it is the hose itself, other times it is actually a bad fitting. To find out, sometimes I place the hose in a pail of water to see where the air is escaping. In general, it has been easier to just build an entirely new hose assembly then mess around trying to fix the one that is leaking.

Using pressurized air to test this has been a real leap forward for me because I used to dread getting the car down on the ground with all the fluids in it—having to fix a leaky hose or two being part of that process. I hated it.

Thankfully, the last two cars I have built have had ZERO leaks when they went on the ground. And that, my friends, is progress!

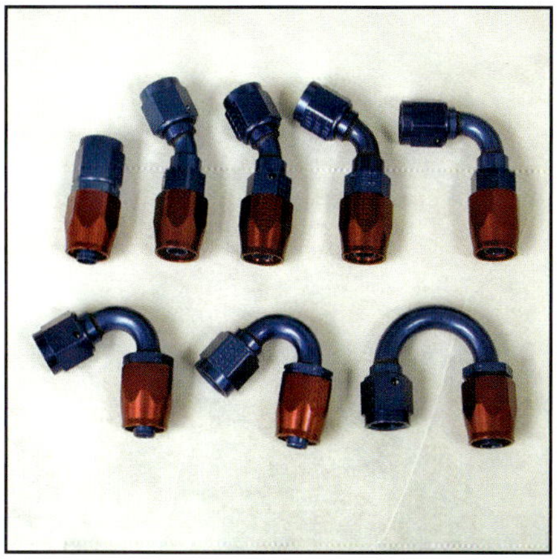

AN hose ends come in a variety of configurations that allow you to route the braided hose as needed. The back row has straight, 30° (2 of these), 45°, 90° hose ends. The front row consists of 120°, 150°, and 180° ends.

One of the most important aspects of plumbing a high performance vehicle is visualizing how the hoses will route. Symmetry is an important aspect of routing hoses, along with ensuring the hose will not contact spinning items like serpentine belts, or come in contact with hot components like exhaust headers.

I lay out a hose through its intended route, allowing for a certain amount of radius to the hose routing so I can match the two hoses routing. This adds symmetery to the engine compartment and keeps the hoses flexible for max durability.

I then tightly wrap the hose with three loops of tape at the desired location and remark my cut line. The tape prevents the steel outer wrap from fraying while cutting.

Cutting the braided hose can be accomplished in many different ways. I have always used a high-speed cutoff wheel.

The Mule was plumbed with Earl's fittings, which feature a cutter style connection (left). The 'cutter' uses the inboard thin tube to guide the rubber lining into the sharp edge that cuts into the hose to create a very reliable seal.

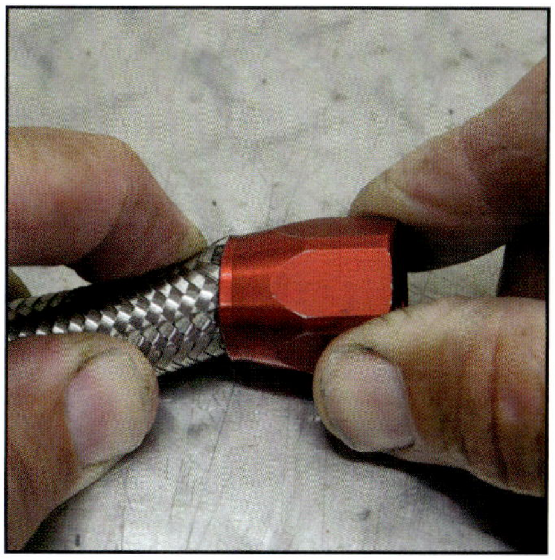

The next step is perhaps the most difficult – assembling the socket over the steel line. Remove the tape from the steel line and carefully slide the socket over the braided steel exterior. Twist the hose counterclockwise into the socket.

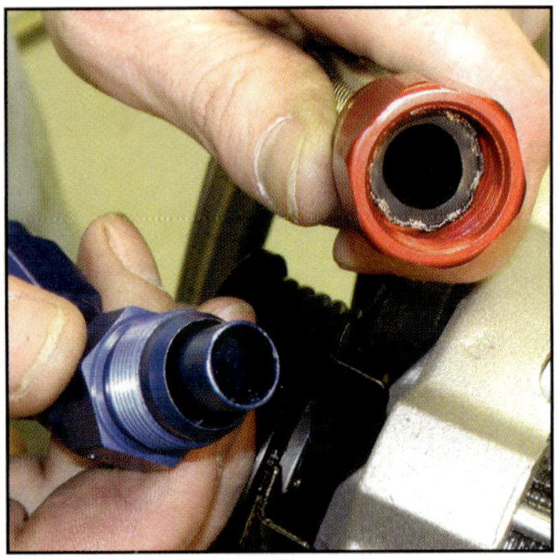

Press the socket over the hose until it is fully seated as shown here to the bottom of the threads in the socket.

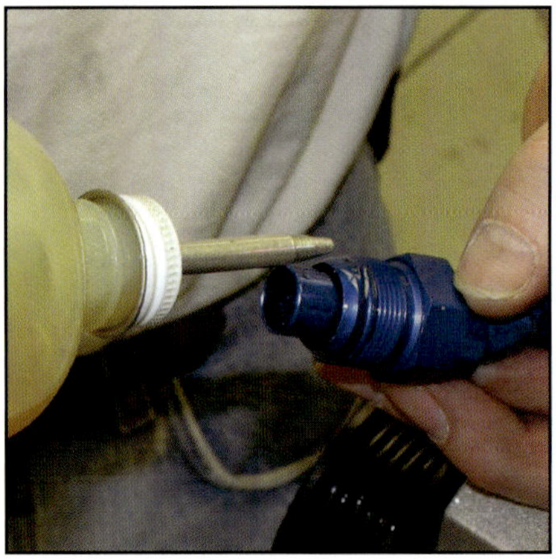

Place oil or water on the cutter and threads to aid in starting the hose end into the socket.

I like to make a light mark on the backside of the hose/fitting mating point so I can check whether the hose 'pushed out' the backside of the fitting during the fitting install process.

These aluminum bench vise AN fitting hold jaws are handbuilt (you can buy them from Aeroquip). I use them to hold the AN fitting collar in place while I use my hands to push the cutter portion of the fitting and hose together, while threading the pieces together.

Once the threads have been handstarted, to avoid cross-threading the pieces, I use the appropriate size end wrench to slowly tighten the hose end into the socket.

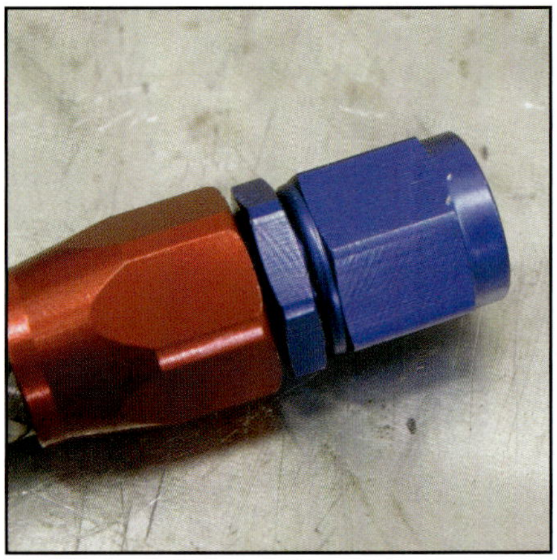

The 'right' way to assemble these fittings is to leave approximately a 1/32-inch (roughly 0.030-0.040-inch) clearance between the socket and the hose end when fully tightened. Then, you can move on to the other end of the hose!

To ensure no grit or dirt remains in the hose after assembly, I like to pour clean solvent through the assembled hose, followed by blowing compressed air through the hose to purge all the solvent and any remaining debris.

The final step is to assemble the hose in its intended location and check for a proper fit. I don't miss too often anymore (but I did a lot in the beginning), but it does happen—in those cases you just chin it and start over.

If you are really conscientious like I am, you will check to make sure the entire cooling system was properly sealed. Rather than test each individual hose, I waited until the entire system was assembled in the car, then used a normal cooling system pressure test tool to test system integrity.

Fuel Tank Lines

To me, the mark of a quality car buildup is when the builder elects to use stainless steel hard line instead of going the easier route of using braided lines front to back. The Mule is an ultimate performance Pro Touring car, so it has a hardline fuel line system.

As a tip, all tubing benders are limited to the radius size of the angle. The larger the tubing, the larger the radius must be to prevent collapsing the tubing. Keep this in mind when planning the route the tubing must take. Also, I recommend you use annealed stainless tubing which will make bending and flaring much easier. You can find annealed stainless tubing from Classic Tube (www.classictube.com).

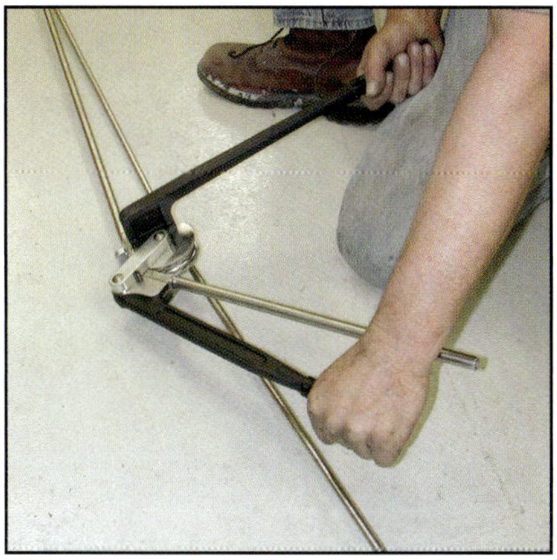

To create the fuel lines from the tank to the engine bay, you'll need ½ and 3/8 inch tube benders, a lot of stainless tubing and some big flaring tools!

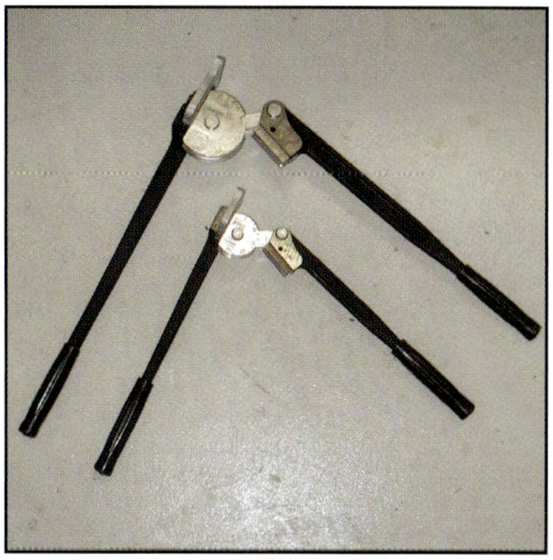

I invested in ½- and 3/8-inch tubing benders to do this job right, thinking this would be the way I'd do it in the future (I have). As a tip, the first bend in a tube length is the easiest, as you don't have to worry about clocking the tubing—from there it is game on!

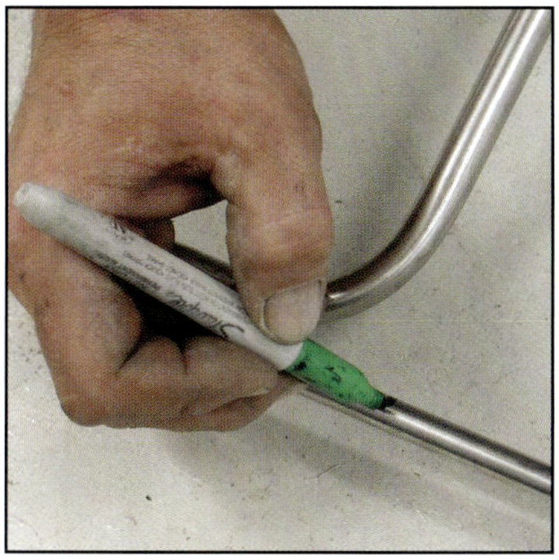

I bent up a ½-inch feed line and a 3/8-inch return line. The ½ inch tube was done first as I wanted to pair the tubes—and the ½ inch tube would have to be bent to a larger radius. A dark Sharpie pen indicated bend centerline and center point of the bend with a 't'.

Here I'm matching the shape of the 3/8 inch tube to the ½ inch tube. Doing this requires some interesting math as the bend radius is different, but I want the tubes to end up pointing the same direction.

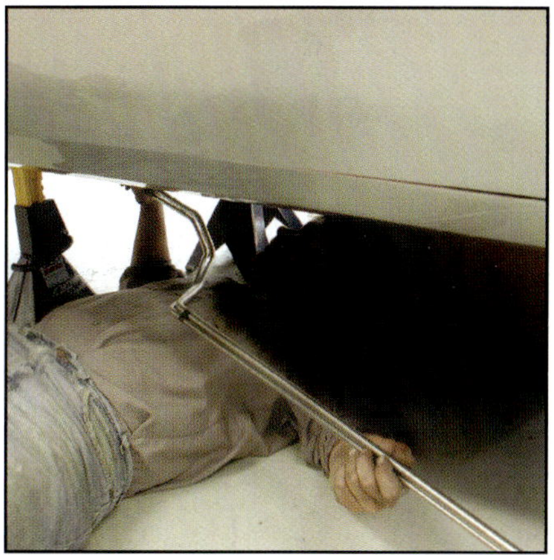

It's always best to double-check your work on the vehicle with each bend to ensure that the tubing will fit the car once completed. Working slowly with plenty of out-front planning will prevent mistakes.

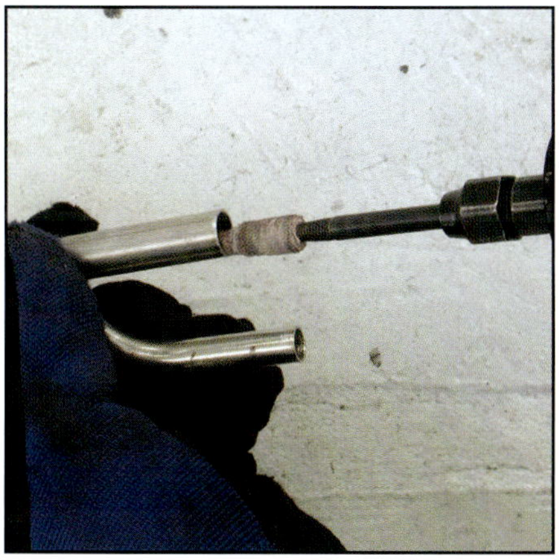

Once the tubing is bent in its final configuration, excess length can be trimmed using a cut-off wheel on a pneumatic die grinder. Then, I use a cartridge roll on a die grinder to remove any burrs in the ID. This helps to insure excellent flares.

Once the tubing is smooth and ready for the flaring tool, remember to install the tube nut and sleeve before crimping the tubing in the flaring tool. The tube nut goes on first, followed by the sleeve as shown.

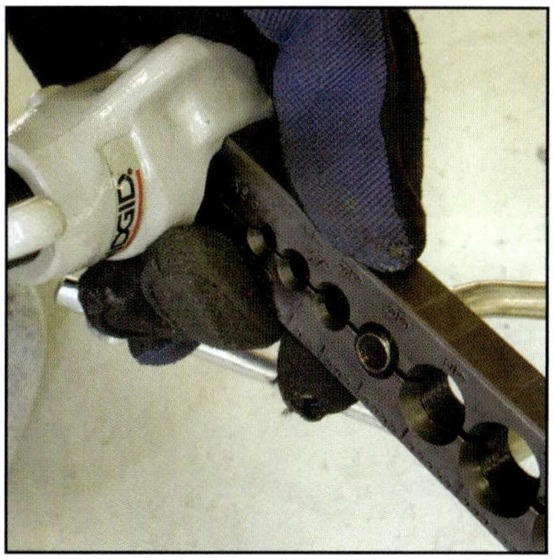

I put a lot of 37° AN flares on tubing with this Rigid AN flaring tool. The process starts by placing the tubing in the tool with its tip just flush with the tool jaws. Extending the tubing too far beyond the tool flat creates a wide flare that may crack.

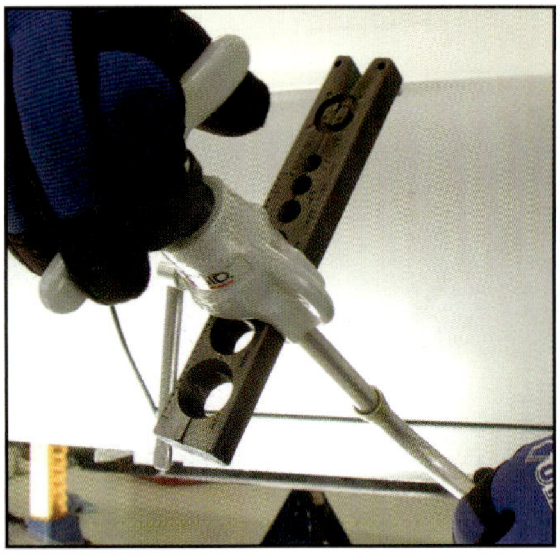

When flaring, turn the top handle on the Rigid tool slowly. This will create a more even flare and prevent possible cracking of the flare. The Rigid tool handle 'freewheels' when sufficient pressure has been exerted on the flare.

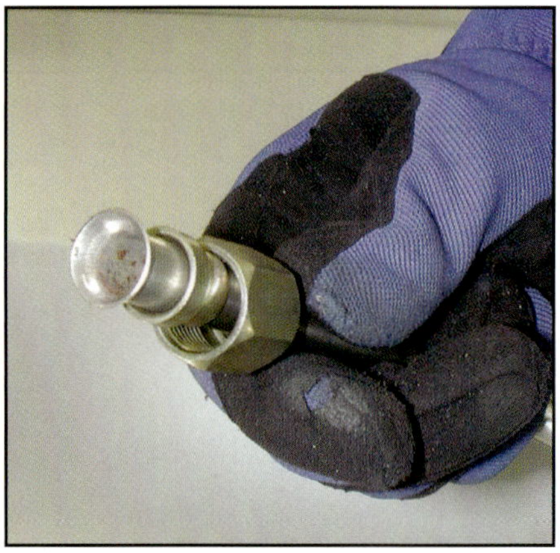

Here's the finished flare with the tubing sleeve and nut in place. This is not the time to discover that you've forgotten to slide them over the tubing!

Since the flared tubing end is female and most hose ends are designed as female, you will need a male-to-male adapter like these to connect the hard line to a flexible AN hose. These male-to-male adapters can also allow changes in hose size up or down depending upon the need.

Here are the feed and return fuel lines positioned next to each other and equipped with the tube nut fittings installed on them. Pretty slick, huh? This took a helluva lot of work, but it tucks up tight under the car and is super clean.

Stainless steel clamps connect the tubes and also mount them flush to the body. Proper layout with an eye toward visual appeal is the key to success when it comes to plumbing hard lines.

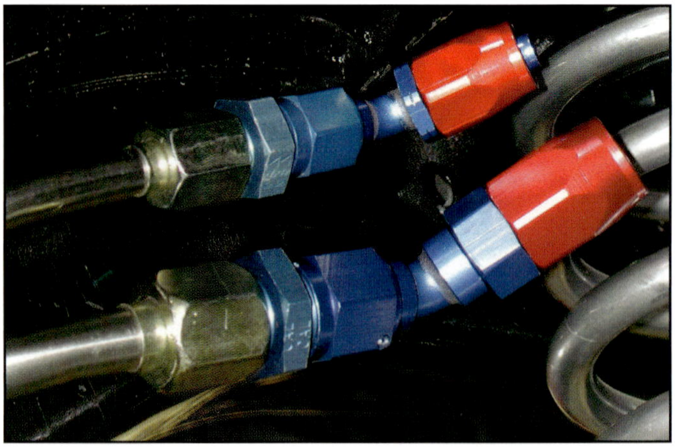

This close-up view shows the final connection to the AN fittings that will connect to the flexible fuel lines (using the male-to-male adapters) that route around the shocks and moving suspension components.

Glass Install

See it all clearly

You are going to be looking out the front and rear glass of your Pro Touring car for a lot of miles, so I suggest you spend some money for new glass and get it installed properly.

If I can find it, I use original glass, but that is getting ridiculously expensive today. The Mule did have original fixed glass installed in it.

You should know that bringing in my glass guy is a real joy, because that tells me I'm getting close to the end of a project!

I have my glass professionally installed, but do the trim install myself. I start out by putting tape down around the perimeter of the glass opening.

I 'tune' the location and angle of the clips that hold the trim in place on the car. These clips need to sit about 1/16 inch off the body to 'grab' the trim and hold it firmly on the car.

I 'black out' the perimeter of the windshield and backglass like they do on production cars to hide the massive bead of butyl that is used to hold the glass in position and seal out any air and water. This gives the glass a clean look around the perimeter.

Sound Deadening for Comfort

Ahhhh, quiet performance

Before you wonder, let me just say 'Yes, I still put sound deadener in all of my cars'. What you see here on the Mule has been repeated on all my cars because the added weight of some Dynamat is not enough for me to suffer with a noisy, hot car interior. If I was building a full-on racecar, it would not have any creature comforts in it…because I know I would only be in it for a short amount of time, and on the racetrack at that, so a little roadnoise wouldn't matter.

I put about 60 pounds of Dynamat sound deadener into the Mule and never regretted it for a minute. This helped the car be incredibly quiet running down the highway and also was part of my work to repel the exhaust heat from entering the cabin.

Rejection Noise

There are a lot of products now available for sound deadening: spray on, roll on, adhesive-backed rubber, etc. They all are excellent options to achieve the same goal: noise rejection. I have used Dynamat for a million years and have found it to be very predictable to install and it works great, which is why I stay with them.

I have used Dynamat on every car I've built in the last 10 years. I start out by laying these sheets in the sun to soften them.

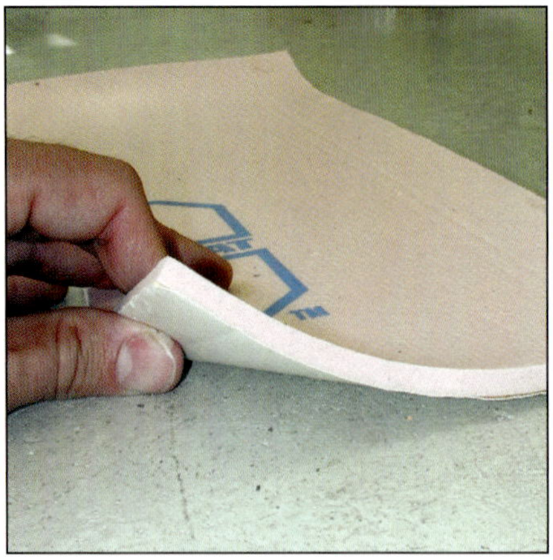

On the Mule, I used this foam heat insulation on top of the sound deadener. It is only about 3/8 inch thick, but the improvement in heat rejection is worth every bit of that.

I like to apply the Dynamat on the interior floor, ceiling, in the doors and anywhere else I can sneak it: I do it by trimming a sheet to fit into an area—making sure to save the cut off pieces as they are just as effective at sound deadening as the big pieces.

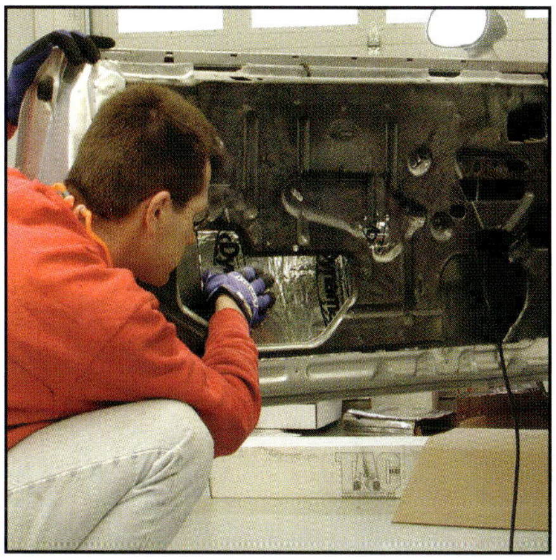

I usually test fit a sheet before removing the plastic covering the adhesive on the back of a sheet. When a sheet fits cleanly, I pull the plastic off about a 2 inch area of adhesive, stick that in place, then pull off the remaining plastic while pushing on the sheet to stick it in place.

Though nobody will ever see it, I work to get an even covering of Dynamat, like you see here under the dash on the backside of the firewall.

Once a sheet has been initially stuck in place, I spend time making sure to push on every surface of the Dynamat to make sure the adhesive is fully seated on the sheetmetal.

Trimming around mounts, pivots or passthroughs (like this one for coolant) requires a little planning—I often make a simple template from cardboard, then transfer that to the Dynamat before installing it.

Trimming, installing and then fully seating the Dynamat to the roof requires some real time and effort—you need to push hard on the Dynamat, but be careful not to get crazy as you can damage the sheetmetal!

This is what the Mule interior looked like after I had installed the Dynamat: notice that there isn't ANY floor sheetmetal visible. I make sure to cover everything! Here, I'm measuring to cut the Dynamat heat insulation foam.

Once I had trimmed/mocked up a piece of insulation foam to the point it was ready for install, I laid it out on a piece of cardboard and sprayed on some adhesive foam.

I then sprayed adhesive where I was going to stick the foam, let it all 'tack up', then installed the foam in the car. I then used a sharp blade to trim the foam around hole openings or other objects.

The areas I insulated included the front footwells and the transmission tunnel. I didn't insulate under the seats or in the rear seat area.

Developing a New System

Or...'How I solved the problem of the protruding, two-speed wiper motor!'

***Note:** This section uses a lot of the elements of a story I wrote about the Mule buildup for Popular Hot Rodding's November 2002 issue because everything I said back then still applies: I still use these tenets in every one of my builds. They have never let me down and I believe you will find more success with your project car(s) by using them. Enjoy, Mark*

There are always ideas bouncing around in my head about systems I want to improve or redesign on my cars. To show you how that process manifests itself into real parts and pieces, I am going to detail one such development process I went through years ago.

Over the years, if there was one thing I really wanted to do differently on my cars, it was the windshield wiper system—yeah, probably not what you expected to hear, but there were three aspects about the stock system that just plain aggrevated me.

The first was that the stock wiper motor stuck way out off the firewall, requiring it to be removed every time I unbolted the driver side valvecover. The second aggrevation was that the system only had the old-timey Off-Low-High settings—no delay feature. The final issue was that the stock wiper motor was just unsightly.

Now, I wouldn't be surprised if you thought the windshield wiper system was a trivial issue to focus on—and usually I would agree. There were a lot of cars before this one where I grumbled about the stock wiper system but bolted it on and focused on 'bigger' issues (like going faster, turning harder and stopping shorter). These other pressing issues would usually trump something like this, but on the Mule, I decided it was going to fixed.

Creating a system from just an idea can be exciting, but also frustrating. Here are some tips on how I develop a new system to maximize the excitement factor.

As you probably know, the system being developed here has been sold by Detroit Speed & Engineering (DSE) for many years now: and is called their 'Selecta-Speed' system. The Selecta-Speed wiper system is based on the prototype system developed here. My friends, Kyle and Stacy Tucker who own DSE, thought this was a product enthusiasts would want, so they created an aftermarket version of this system. DSE's engineering team developed a very slick wiring harness, custom controlling switch and a control knob that matches the Camaro knobs. It has worked out so well, DSE developed the Selecta-Speed system for all Camaros from '67 to 81', the '64-72 Chevelles and the '63 to '74 Novas.

My Prototyping Process

1. Identify problems with the current system
2. Determine why these problems exist and develop a list of goals for a new system
3. Research current technology to determine if anything exists to solve the problem (This falls under the 'Why reinvent the wheel when you don't have to?')
4. If components/system(s) exist, modify the current technology to fit your needs and skip to Step 7
5. If systems don't exist, find a specialist in that technology and use them as a resource to develop a new system (I am the first to admit I don't know it all and don't want to spend the time reinventing something if the learning curve has already been climbed by someone else)
6. Develop a basic system to determine if a fix is possible
7. Build an operational system for the vehicle
8. Refine the system via on-vehicle testing

Why Prototype a System?

I'm always asked, 'How do you know when to build a new, one-off system?' Well, it's a good question and my answer is really simple—I do the work if I think my 'new' system will:

-offer more performance,
-be lighter,
-be simpler,
-be smaller,
-be easier to service/use/adjust,
-look better,
-be more cost efficient, or
-just be better *(sometimes that comes down to me just feeling better about it on the car).*

Getting Started

Once I have decided to jump in on building a system, I look around to see if there is a production or aftermarket system available that will fulfill my needs. For instance, I refuse to have a hot rod that doesn't have air conditioning that can practically freeze me out of the car. Thankfully, the Vintage Air A/C system does just that, so I install it and move on. If there wasn't an A/C system that did the job, I'd build one. When starting on a prototype build, my goal is to create a system that improves or corrects more than one of these deficiencies (listed above as improvements). I either write down what I don't like and would like to improve, or if it's a short list (like with the wiper system), I just keep these thoughts in my head as I get to work.

Once I know what I want, and that there's nothing available to bolt in, I start looking at modifying original equipment manufacturer (OEM) components to achieve my goal. In the case of the wiper setup, I went to the junkyard and pulled many different wiper motors off everything from Hondas, to semi trucks, mocking them up to see if they would work on the firewall of the Mule. After a few iterations of this, I settled on a wiper motor off a late model American car.

I started with import vehicles because I thought those would be the most compact. While they were small, they were pretty ugly, and I had to cut them up a ton to get them to fit on the Camaro firewall. One of my engineering co-workers told me the big-trucks had compact, clean-looking wiper systems, but I couldn't find any that fit cleanly into the small area on the firewall. The late-model wiper motor I chose is by far one of the most compact, clean looking wiper setups I found that offered all the features one would want on a modern system. After getting it bolted up to the firewall using a custom-built adapter plate, I reverse-engineered the wiring system to get it running.

The final step was to cut the Camaro and late-model motor wiper arms in half so I could weld them together. This made it easy to mate the differently splined late-model wiper motor to the early Camaro armature.

By far, the hardest part of creating this system was finding the wiper motor that had the features, compactness & appearance I wanted—once that was in hand, the rest was just tying up the details. This took me a long time, consumed a lot of energy and seemed like a lost cause on a few days…I never said this type of work was easy, quick or a sure thing. I run through a lot of deadends before making progress. To me, that is what Engineering is: problem-solving via real life trial and error (and yes, there is a lot of computer modeling done today to speed up problem-solving—but where do you think the models for the computers came from? Yup, real life trial and error).

Building from Scratch

I was lucky with the wiper system because I didn't have to build a 'delay' controller to run the wiper motor. If I hadn't been able to find something with that feature built into it, that's exactly what I would have done. It would have taken a long time, been painful and probably not have worked as nicely or been as durable (OEM companies spend millions developing a component or system like this), but it would have got the job done.

I recommend you search high and low for current technology before deciding to build a complete system. I have found using even the simplest OEM components as a starting point is an excellent way to build a durable, cost-effective system that performs with the desired functionality.

Luckily, since Detroit Speed and Engineering offers a few versions of this delay wiper system for multiple body styles, you probably won't need to go through the many hours of research, development, fabrication and refinement that I, and then DSE, did to have intermittent windshield wipers on your Camaro, Chevelle or Nova. But, if you decide there is something else you must have and it's not available, at least now you'll have any idea how to do it.

Once I found the 'right' late model delay-equipped wiper motor, the improvements became obvious: like the low-profile nature of the motor. The first time I had to remove the valvecover with my new system in place made me smile.

To make it easy to create the aluminum adapter plate for the late model wiper motor, I cut off the stock mounting plate that comes on the factory first-Gen Camaro wiper motor (in right hand) and used it as a template.

The ¼ inch aluminum billet adapter plate was cut out with a bandsaw and the bolt holes lined up via the factory bracket. I attached it to the new wiper motor with countersunk Allen bolts.

When I first bolted the late model wiper motor up to the Mule's firewall, I thought it looked great, but wasn't sure whether it would 'sweep' the wipers correctly or how I'd wire it up—but I figured those were problems I could work through once I got the motor right.

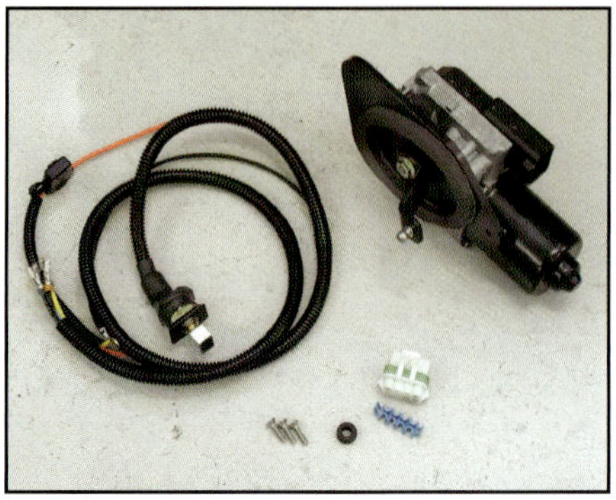

Luckily, for anyone wanting this on their Camaro/Chevelle/Nova, Detroit Speed & Engineering (DSE) offers their Selecta-Speed delay wiper system as a bolt in kit. The DSE kit comes with the bolt-on wiper motor, a wiring harness and a custom switch. This has got to be one of the easiest-to-install aftermarket products that actually works and looks good.

Here was one of the first issues I had to solve once I got the late model wiper motor installed on the car: the Camaro wiper motor has a 'half-moon' shaft and the late model motor is fully splined.

I solved wiper motor shaft mismatch issue by cutting both the Camaro and late model wiper motor arms and welding them together (they are shown before I cut them). DSE solved this issue by creating a custom arm from billet that looks good and bolts on easily for their Selecta-Speed system.

The new wiper motor and armature were bolted up to the car. After discussing the wiring schematics with some knowledgeable sources, I wired my Optima battery and multistep switch to a small circuit board to test the system. The motor rotated at different speeds, but I couldn't tell if the arms were sweeping the arc…

I taped some welding wire to the wiper arm spuds to determine what the arcs were. I was happily surprised to see the arms sweep essentially the same arc as the stock wiper motor! What can I say, sometimes you get lucky…

I always make sure to grab the mating electrical connector for whatever I'm pulling out of the salvage yard. This way I know I can connect wiring to it when needed.

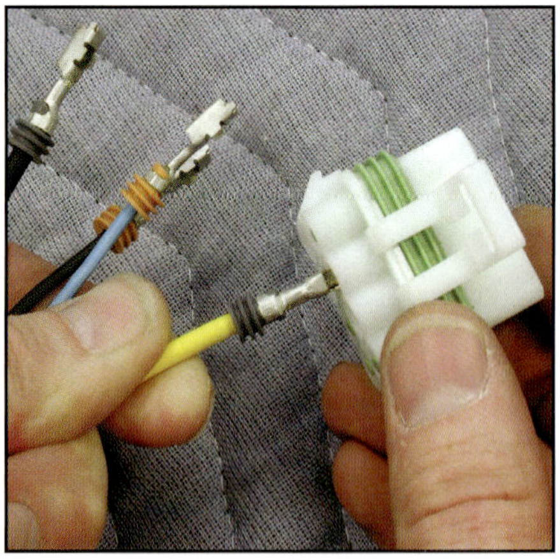

Places like Terminal Supply in Troy, MI can sell you the correct terminals so you can rewire the connector for your application.

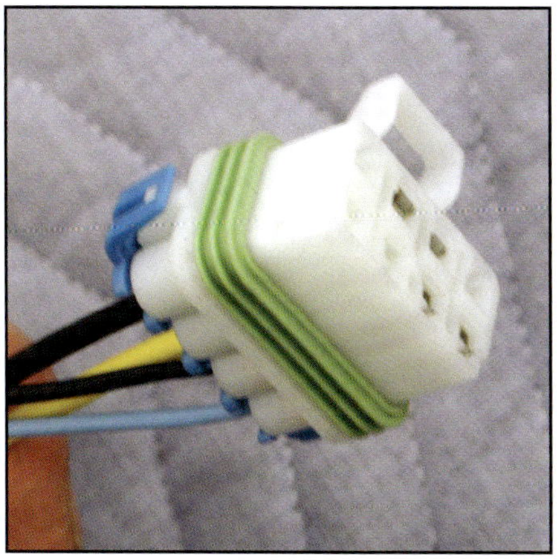

This type of OE electrical connectors are great because they have a seal built into them, so they keep dirt and moisture out of the connection. Nice.

So this is the mechanical end of my prototype delay wiper system for the Mule before I installed it. Looks like it was meant to be, doesn't it?

To install the wiper motor on an already built 1st Gen Camaro, removing the cowl cover is required. To do this, the wiper arms need to be removed, the cowl bolts removed and the cowl carefully finessed off the body (patience is required with every step of this).

Once the stock wiper motor is removed (unplug the wiring, unbolt the armature inside the cowl, unbolt the stock wiper motor), you can bolt up the DSE Selecta-Speed motor (in this case, I am bolting on my proto-wiper motor).

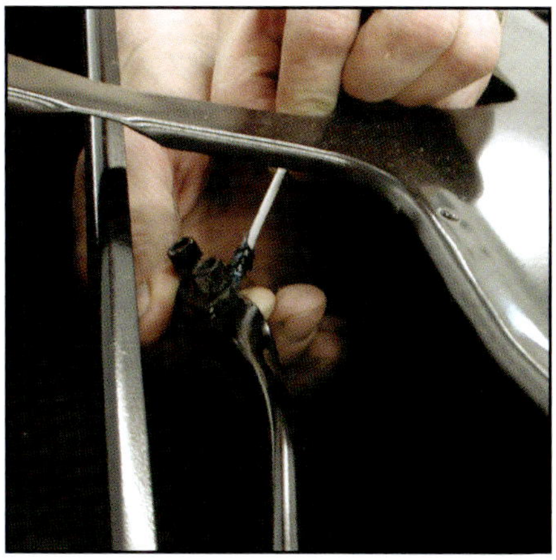

I always clean out the pivot arm mounting point as it fills up with gunk that limits movement. Solvent-soaked cotton swabs are good to clean it, then liberally apply wheel bearing grease to this area before assembling the system: this should help the system operate smoothly.

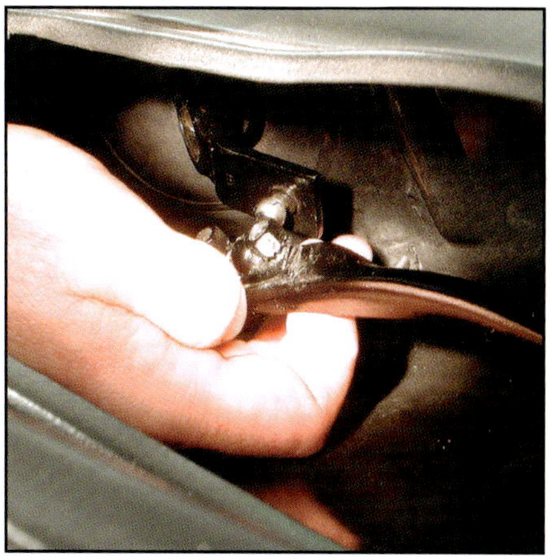

Here I am connecting the stock wiper armature to the delay wiper motor arm (actually the combination of the stock Camaro arm and the late-model arm). This mating point is heavily greased up, too. DSE has built a specific armature for this swap that bolts right up.

Tightening up the two 7/16 inch nuts requires some contorting of your arm to get into the cowl space, but it goes easily. I suggest leaving the cowl cover off until you get the wiper motor wired up and running properly—just in case!

I have always enjoyed using the delay wipers on my Pro Touring cars ever since doing the original package on the Mule (I have bolted on DSE Selecta-Speed systems since that one)—and now you know how I create a new and improved system.

Wiring it all up

How to build durable, good looking and easily serviceable wiring

Note: This section uses some elements from a story I wrote about the Mule buildup for Popular Hot Rodding's December 2002 issue because what I said back then still applies to how I wire my project cars today—and probably will for the rest of my life! I hope it all helps you. Mark

If there is one task in a hot rod buildup that should be labeled 'thankless', it should be wiring the electrical system. Wiring is an extremely detail-oriented, time-consuming, tedious job that requires many special components and tools to accomplish successfully—but nobody ever sees it or compliments on how nice your wiring is. They only say something if the wiring system DOESN'T work properly.

I would really prefer not to do this part of the project, but know it would be very expensive and take a long time if I shopped it out. I am usually excited about getting closer to having the car done when it comes time for wiring, so instead of parking it in a wiring shop for who knows how long, I do it myself.

Having wired most of my projects, I have refined my method to create an electrical system and offer here what I've learned from those years of learning.

Wiring 101

As a qualifier, I'm going to focus on the overall methodology of the job, so if you're just starting out in the world of wiring, I'd suggest you read up on electricity, vehicle wiring systems and component requirements for power before beginning to wire your vehicle. This research will give you the fundamentals to understand the schematics that come with the aftermarket wiring systems (either the Painless Products system used here or an original equipment manufacturer (OEM) replacement system like those available from reproduction outlets like Year One). I'll explain some of the basics, but really, you'll need to learn the fundamentals offline before jumping into a project like this.

Wiring a car. I really don't like doing it. But, it is one of those steps that needs to be done…and done right. Here is how I do it—and yes, my garage usually looks like this when I am in the middle of wiring a car (the small bins on left hold connectors, wire, pins, etc and big bins hold old wiring harnesses and late-model parts and pieces).

What I want from the wiring system is durability, performance, ease of maintenance and repair while ending up with a system that is simple and light. Once done, the wiring system should never be a topic for discussion. If it is, more than likely that means it isn't working properly. While some car builders spend time and effort getting all the wires 'straight' in the bundles, I'm more concerned with keeping the wiring system functional and simple. Let's face it, most people will never, ever see the wiring system, but the system does need to work properly and be simple.

The wiring system will be only as good as the components you use to build it. Along those lines, I like to start with a system like the 1969 Camaro Painless Performance Products OEM replacement system. It installs in the stock location, but has the capability for double the amount of circuits of the stock wiring.

I then use a combination of OEM connectors and Packard Electric connectors, GXL wire, Year One banding tape, Year One factory front and rear lighting harnesses, plus Moroso and military-spec bulkhead connectors as I modify the Painless system to fit my needs. I recommend having a lot of these components on hand, and in big quantity as the variety of materials needed while wiring a car is staggering. If you don't have these products in the shop, you will spend most of your time running around, or mail ordering them—and this can consume an insane amount of time. For me, this is the major reason wiring a car is so challenging: knowing what you'll need and having it in quantity (and knowing that costs real $$$).

As a disclaimer, you should know I have an entire shelving unit in my shop filled with plastic bins loaded with wire, connectors and other wiring products that I have bought or acquired from junkyards/project cars over the years. Every time I wire a car, all of these bins come out and I use something in just about every one of them—it still surprises me how many different products are needed: wire, terminals (which are also called 'pins'), connectors, solder, tape, shrink tubing, zipties and more. Also, I always keep the original wiring harnesses off my cars in case I need a connector or some wire… and they are great for understanding how a harness was run.

The Process

As with anything, there are million ways to wire a vehicle, but here's my system. Once I have the materials laid out on packing blankets on the ground (the products are mostly in plastic bins) and the tools on the workbench, I start at the front and rear of the vehicle and wire to the middle.

None of this happens until all the components that either generate or require power are mounted in the vehicle. As I've said, I like to use the Painless '69 Camaro wiring kit as the foundation for the system. It has 18 circuits versus the stock 9 circuits, which means it can easily power high-amperage components like big radiator fans, electric fuel pumps, air conditioning systems and stereo amps.

While I often hear pros talk about how they start with the most difficult area of a system to get it out of the way first, I can't do that when doing wiring. This job is hard to get up for, so I start with the easiest area: the rear lighting harness.

On the Mule, I had a reproduction Year One rear harness connected to all the exterior and interior lights throughout the trunk area in about five minutes. This type of success feels really good and gets me fired up to finish the rest of the car.

As another way to keep me exciting and moving forward, I often will take a few minutes to temporariliy power up a leg I have just finished. I do this after referring to the Year One schematics (which usually have the same color coding as the OEM system) so I can temporarily wire in my Optima battery, with a fused link, to 'power up' elements, like the lights, to see how they look. This makes a big difference in my excitement level and ability to keep moving forward on this big job: knowing that another area on the vehicle is working properly means I am that much closer to having it all completed.

Powering up separate legs is also a great way to check your schematic reading and assembly procedures. You definitely don't want to keep going forward if a subsystem is wired incorrectly. Along those lines, I recommend you get the stock vehicle wiring schematics and compare them wire-for-wire to the Painless kit before you start the job—making notes on where the system will vary from the stock system. This way, you'll have a roadmap before you make one wire loom—and long before you have a wiring problem.

Wiring your own vehicle has its advantages. If you should ever have a problem or want to add another circuit, you'll know how the system was created and what to do to get back on the road. I have used this knowledge many times—it is those rare moments when I think of the job I did and am thankful I did the wiring! Keep good notes on your custom or added in circuits. It will help you troubleshoot later.

9 Tips For Top Electrical System Performance

1. Spend the money for the right tools to crimp and solder each type of connector.
2. Use quality terminals and connectors!
3. Once each system is finished, test it with power so you feel the success and gain confidence in your progress (use a fuse and fused link in case you've got it wired wrong).
4. If you have an electric cooling fan, fuel pump and other big-power/high-load components, get a 140-amp alternator and use 8-guage wire to the starter/fuse block (not the 12-guage that comes with many systems) to handle the stress of running these powered components
5. When building the front lighting harness, leave the grille and radiator out of the car so the harness can be easily slid in and out of the engine bay through the radiator core support opening
6. Use non-adhesive banding tape from Year One and bind in the stock clips instead of putting the wire in the split loom
7. Lay out the entire harness on top of packing blankets on the floor and loosely ziptie the wires together, then test fit it in the car. This way you can see what the system looks like installed but changes can easily be made.
8. Use a high-quality bulkhead fitting to run the 2-guage main battery line through the firewall as this piece must always work in a high heat, moist and dirty area (the Mule has a Moroso bulkhead fitting).
9. Be careful when and how you mount an electronic ignition box. The RF (Radio Frequency) that comes off the coil wires can play havoc with your other electrical equipment (I wired up a shielded wire from my MSD unit to the coil to fix the feedback through the stereo).

Stielow's Wiring Process

1. Install the Year One rear lighting harness
2. Install the Year One front lighting harness
3. Build the engine control electrical harness
4. Mock up the Painless '69 Camaro 18-circuit bulkhead system and begin modifying it to fit my needs
5. Install main wiring harness.

Adhesive Lined Shrink Tubing

Tip: I use adhesive-lined shrink tubing because I like the idea of the shrink tubing gripping the wire on the terminal and providing a near watertight seal. The only problem with this adhesive-lined shrink tubing is that it can't be cut off and the terminal modified once in place. It's a final deal. You can still work with this, but just know that you'll need to cut off the end and start over if you want to make a change. I find that acceptable as the end is so much more robust with this type of shrink tubing (vs no shrink tubing or regular shrink tubing).

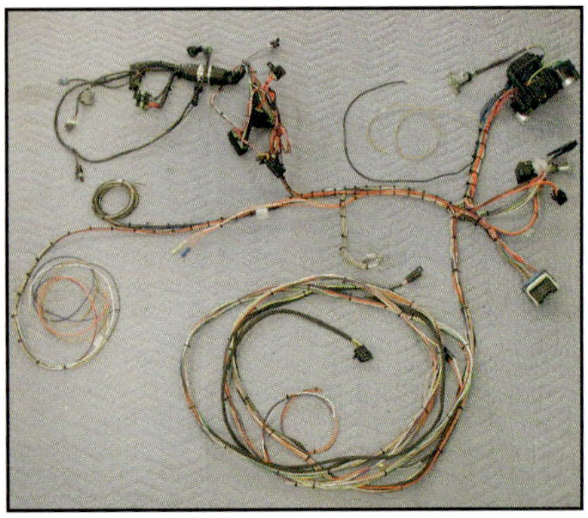

This is what the main wiring harness looked like on the Mule just before I wrapped it in wiring loom tape and installed it in the car. The small harness in the upper left is for the engine controller (I will show how this was built later in this section).

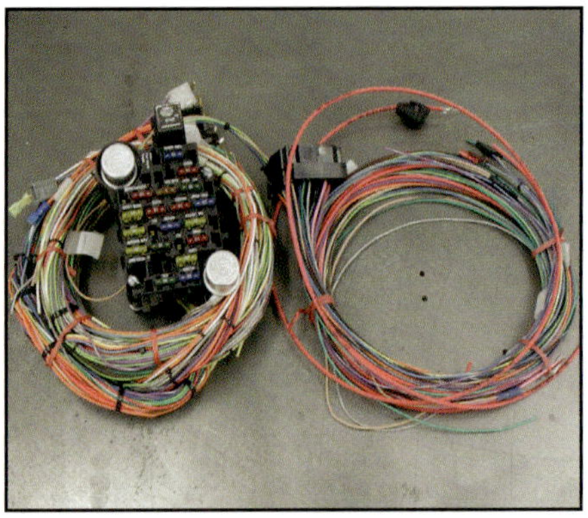

I like to start with an 18 circuit 1969 Camaro replacement wiring harness kit from Painless Performance Products. I have used a Painless kit on my cars for years because they have plenty of circuits for all the components I need to power, install easily, provide durable service and look 'right' on the car.

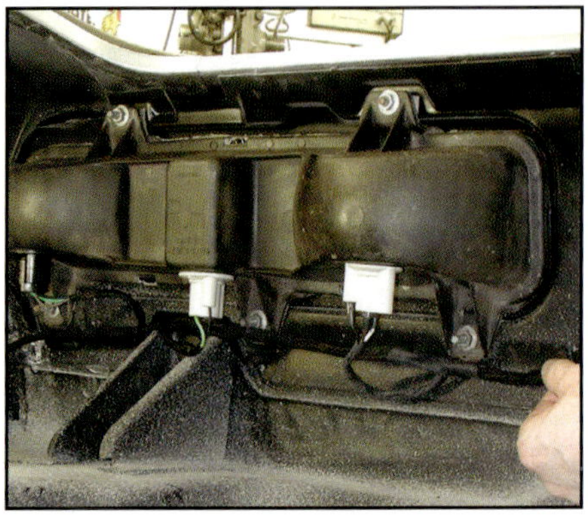

The easiest wiring harness to install is the rear lighting harness—in this case a Year One repro unit—and is why I start with it. This gets me going on the project and also gives me a chance at early success…which I need when doing the wiring!

Installing an Eyelit Terminal

I use the following process when installing an 'old style' eyelit terminal on some wire. I have found this method to be able to withstand vibration, dirt, heat and other negative forces yet still convey the power.

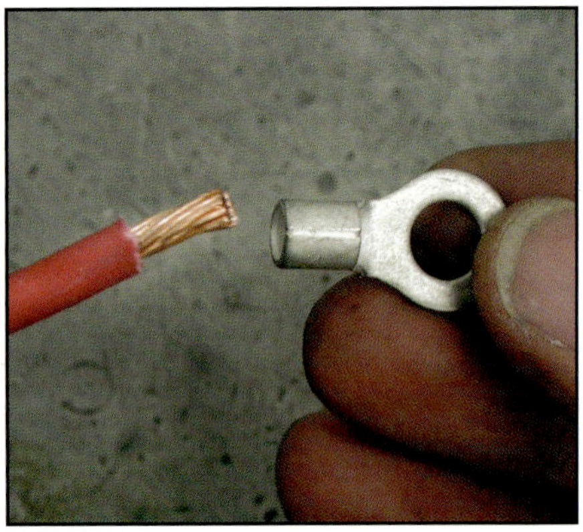

First, I slide a ¼ inch diameter, ¾ inch long piece of shrink tubing onto the wire (REMEMBER THIS!). Then, I use a Blue Point wire stripping tool to remove 3/16 inches of insulation from the 18 guage wire. I then install the terminal end on the wires in prep for crimping.

I position the terminal/wire combination in the jaws of a Packard Electric crimping tool and clamp down as shown. This closes the terminal tangs onto the wire. You can buy these larger size crimpers on Amazon.com.

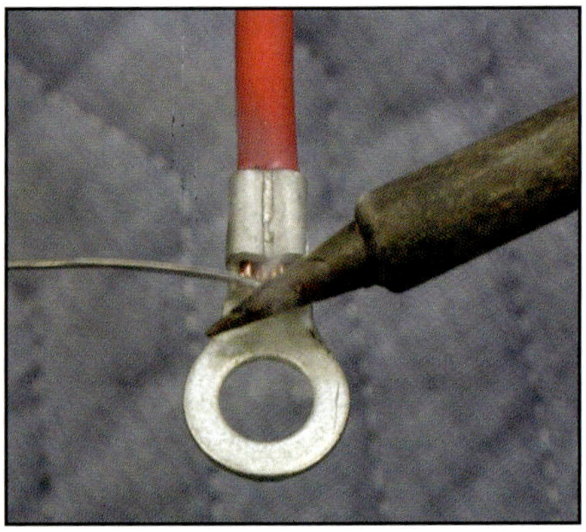

As an extra level of security, I solder the wire to the terminal using 40/60 solder and a Weller soldering iron—drawing the molten solder into the wire/terminal mating point.

Here is a properly crimped and soldered terminal. Notice how the two tangs of the crimped terminal grab the wire while the backside is still curved (instead of being mashed flat). To get this, you need to use the correct crimpers.

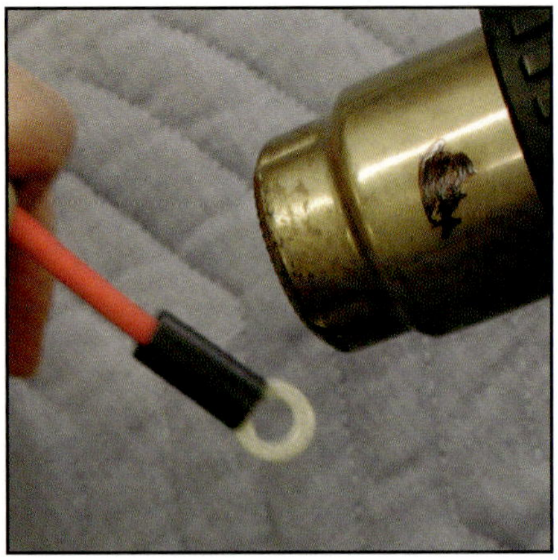

The next step is to slide the heat-shrink tubing over the terminal ends (you did put it on before crimping on the terminal, right!?) and apply even heat from a heat gun to shrink the tubing on.

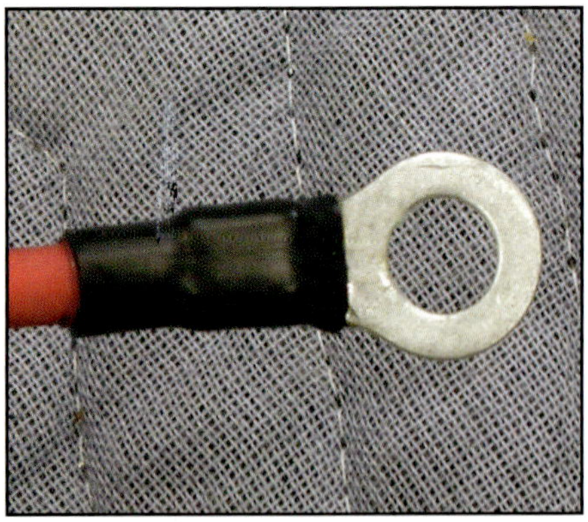

The installed eyelit connector looks good and will provide a long life delivering electricity. Notice how the shrink tubing completely covers the crimped tangs and an equal length of the wire. This is how I determine the length of shrink tubing to use. Also, you can just barely see a small bead of adhesive that has boiled out from under the shrink tubing where it meets the wire to provide the watertight seal.

Installing Terminals and Connectors

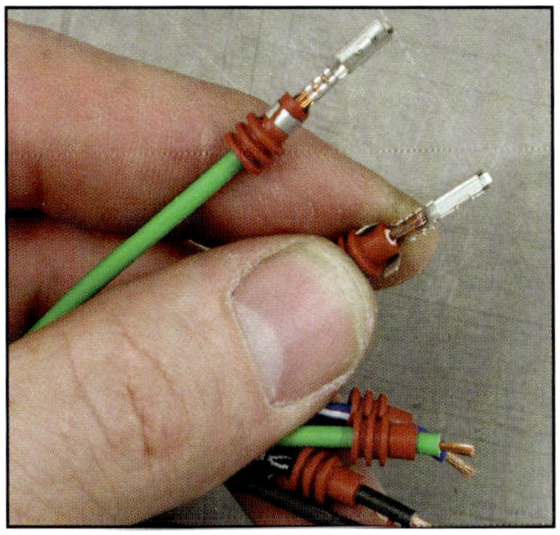

These male Packard wiring pins are provided loose in the Painless kit. They require special Packard crimping pliers to properly install. Notice the little seal I have slid on the wire before stripping ¼ inch of insulation and positioning the pin for crimping.

To crimp on these pins, I use these stepped crimping pliers, placing the open end of the crimp-on section of the pin in the deepest throat, pointy jaw form. I close the pliers until the tangs are folded over on themselves or the pliers bottom out—in which case I move to the next most shallow section located on the pliers to finish folding the tangs in completely.

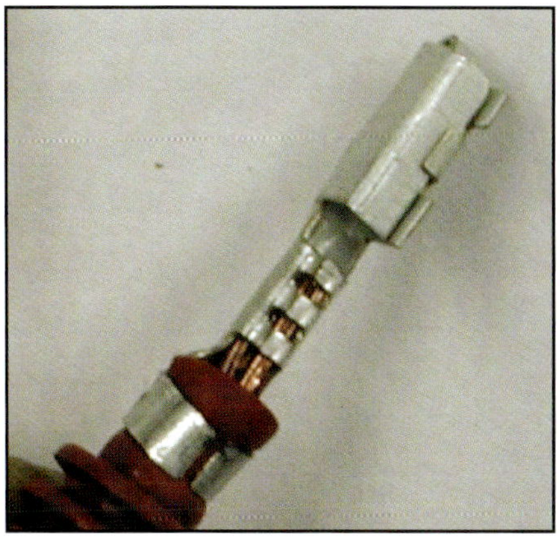

Here is what the pin should look like when it is completely crimped on the wire. Notice how the grommet is crimped onto the pin—this is accomplished with the same pliers, but using the larger round jaw sections to complete the crimp.

The pins with the wires crimped on them are then slid into the multi-pin connectors in positions that are either called for in the factory wiring schematics or in a schematic that I create. This is where you'll need some capability to follow a schematic or create your own.

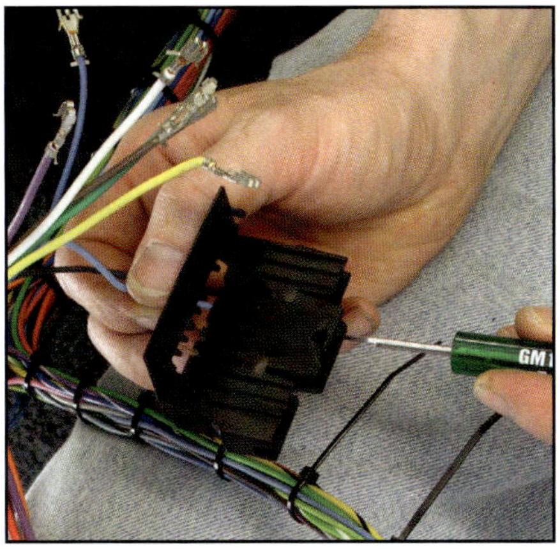

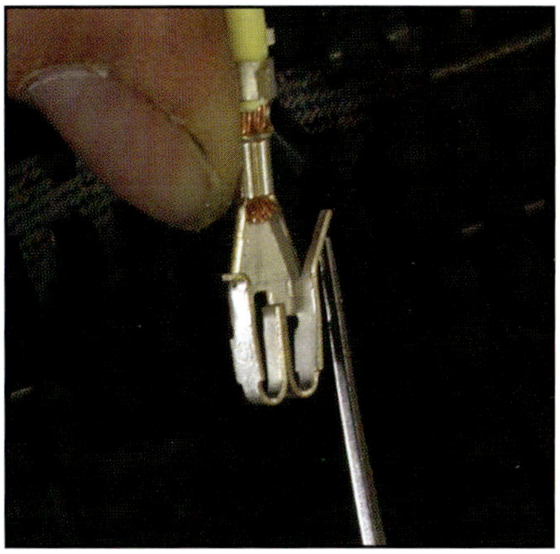

Regarding the location of the pins, on the larger connectors, sometimes I change the orientation of the wires to fit my needs. To push the female pin out of the fuseblock, a special wiring tool (in this case a GM tool, pn 12054429) is used to push the lock tang (photo on right) in so the pin can be pulled out the back of the fuseblock.

Weatherpack multi-pin connectors, like this two-pin connector, require the wires be threaded through the connector and connector seal (on left) before installing the pins on the wires—but the actual crimping of the pin on the wire is similar to the Packard connector.

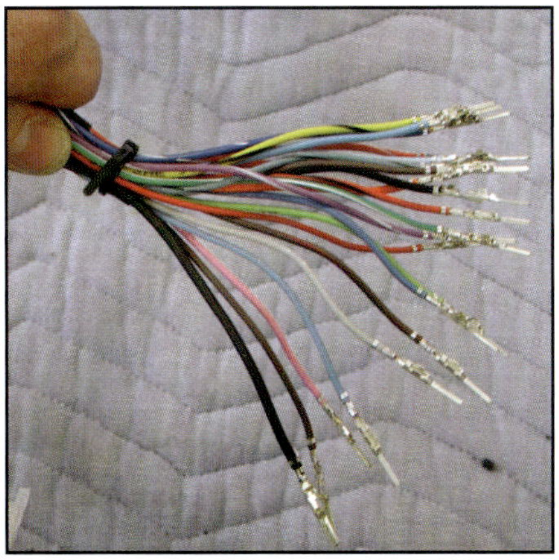

Here is a bundle of male pins ready for installation on connectors. Now you know why a written plan of the schematics is needed—this would be a mess without a good roadmap of where everything goes!

As I'm getting closer to having the body wiring harnesses finalized, I continue to 'pin out' the main fuseblock to get everything lined up.

Once the fuseblock pin locations are locked down, I set the male pin locations on the connectors by pushing them into the multi-pin connectors.

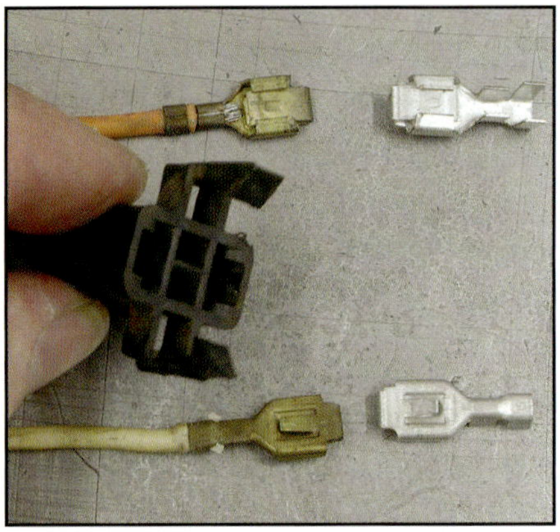

The old-style 'blade' pins are used on some of the period-specific components—I save the good connectors off the old wiring harnesses for just this purpose and install new pins on the wiring.

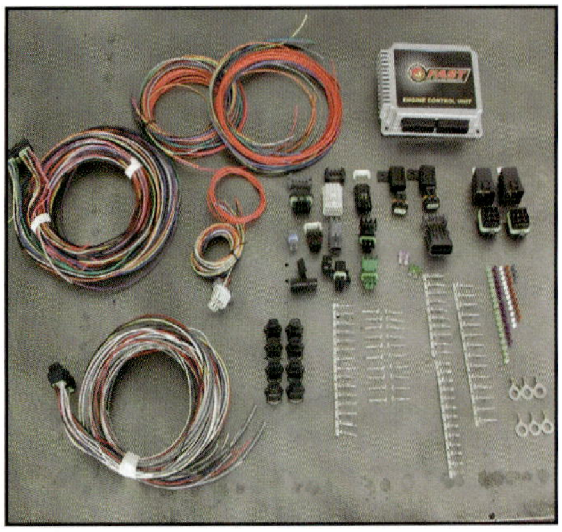

The Mule utilized a F.A.S.T Engine Controller kit and wiring harness that worked great at getting the engine to start on the key, idle smoothly, have great part throttle control—and obviously, excellent Wide Open Throttle (WOT) power.

Wiring the Engine

The late, great Mike Browne, of Wires & Pliers fame, did the engine harness on the Mule, and there is plenty to learn from how he did what he did (and he let us shoot photos of how he did it so I could show you).

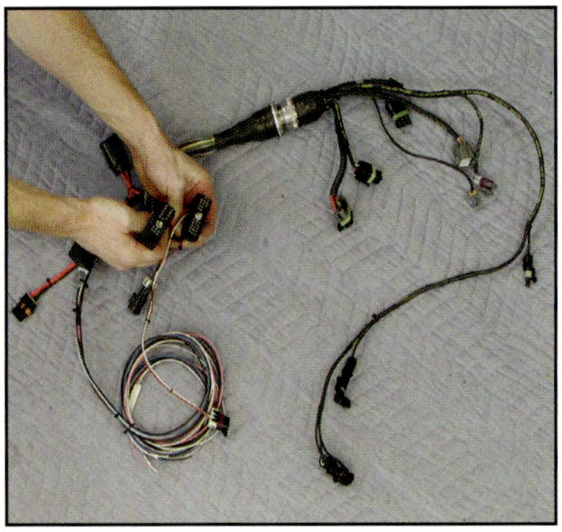

I wanted to be able to remove the engine quickly, so I had Browne wire in a 41-pin military-spec quick disconnect for the engine harness. The connectors in my hands attach to the F.A.S.T. controller.

Fuel Injector Wiring

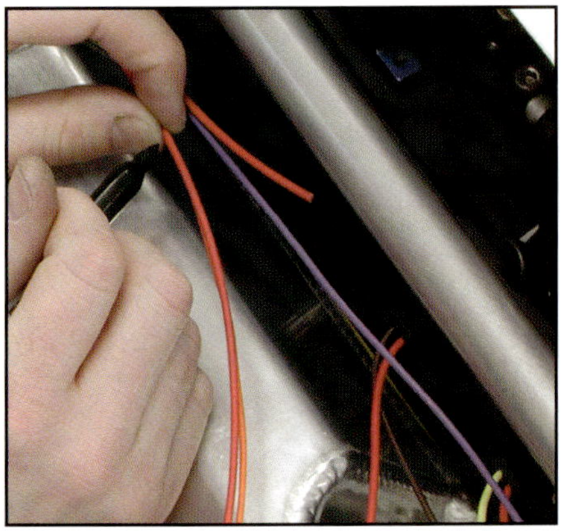

Before Browne cut a wire, he would mark the length with a Sharpie pen so he could look everything over before making a cut.

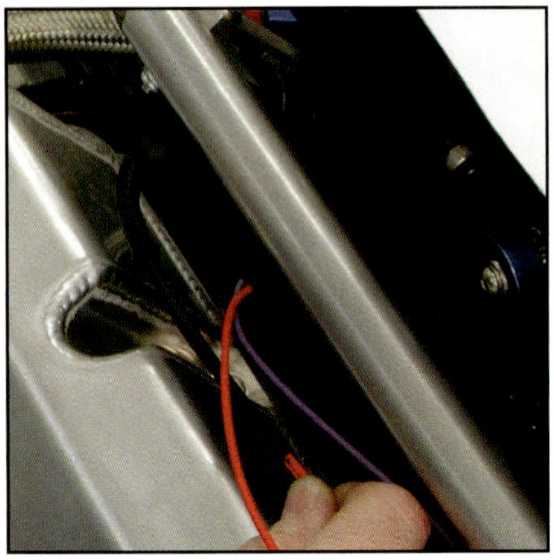

Each electronic fuel injector has two wires to actuate the solenoid in the injector (and allow fuel into the intake port). Browne installed black rubber wiring connector covers over the F.A.S.T. connectors and also used shrink tubing on the wires.

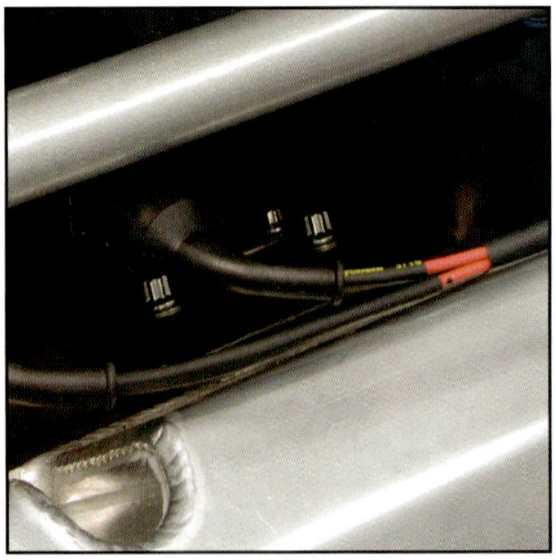

Here is another view of the fuel injector wiring—looks as good as it worked: perfect!

A fuel injector wiring loom was built using various diameters of shrink tubing over the wires.

Weatherpak Connector Buildup

All late-model cars use 'Weatherpak' fittings—which is good as they are more durable than the Packard or old-car blade-style connectors. To put one together is a little different than the Packard connectors: it starts by sliding the connector grommet and connector on the wires.

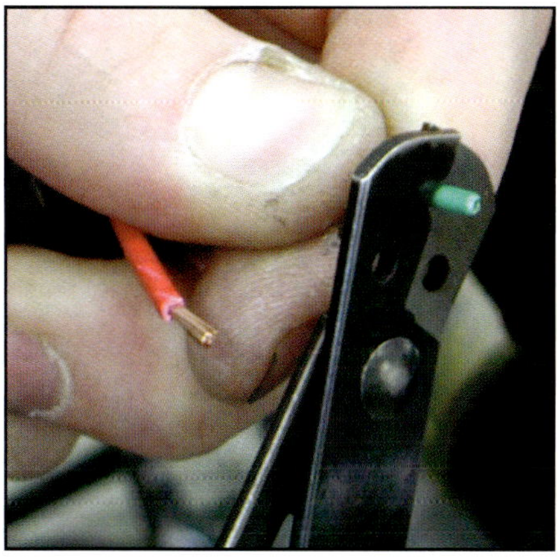

The next step is to remove ¼ inch of insulation off the wires. The pros use these to strip the wires, but I like to use the stripper pliers that have slots in them for specific wire sizes as it is less likely I'll cut into the wires when stripping the insulation.

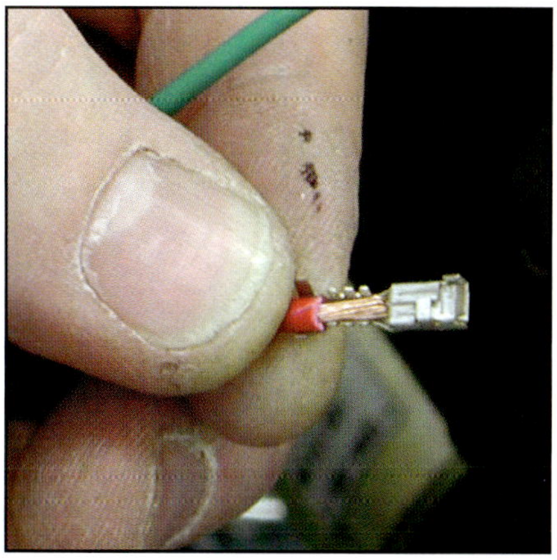

Here is the stripped end nested in the end of the Weatherpak pin before the small tangs are crimped on to the wire—notice how the bigger tang at the end sits over the insulation, ready to be crimped on, too.

Using the multiple depth jaw segments of the pin crimping pliers, the tangs are compressed until they are bottomed out on the wires. This requires some practice to do properly, so expect to do a few before you get the tangs to close evenly in the jaws.

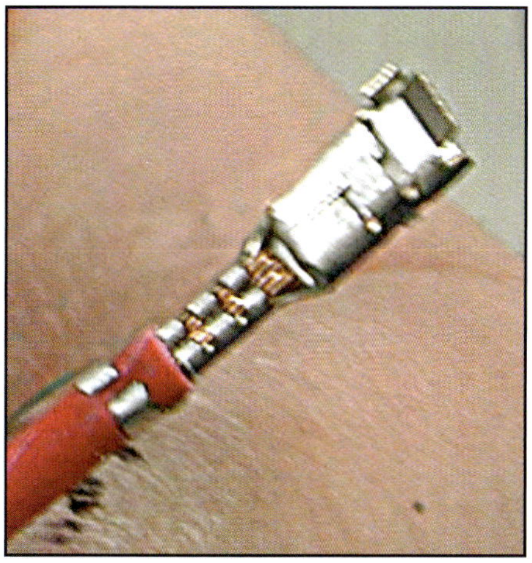

Here is what the connector should look like if you've completed the crimping process cleanly. This is harder than you'd guess to achieve! This is a good crimp.

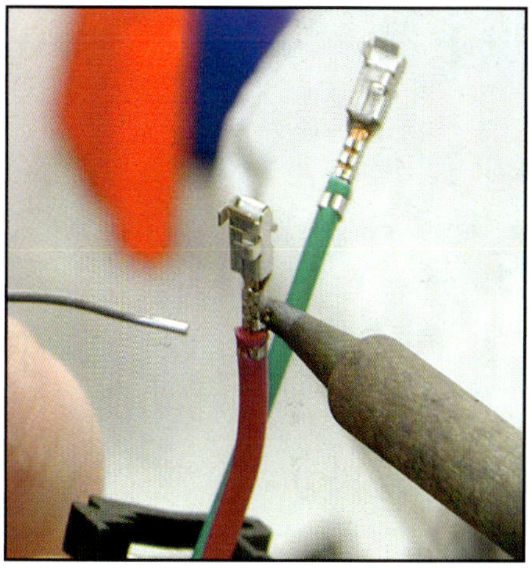

As added insurance, the wire to pin crimp can be heated up and some solder is 'sucked' into this area to further lock the wire onto the pin.

This shows a pull to seat terminal for an injector connector.

Minimizing Wiring

The wiring for the engine was positioned in the intake valley to minimize its visual impact, and black zipties and shrink tubing used to further hide the wires.

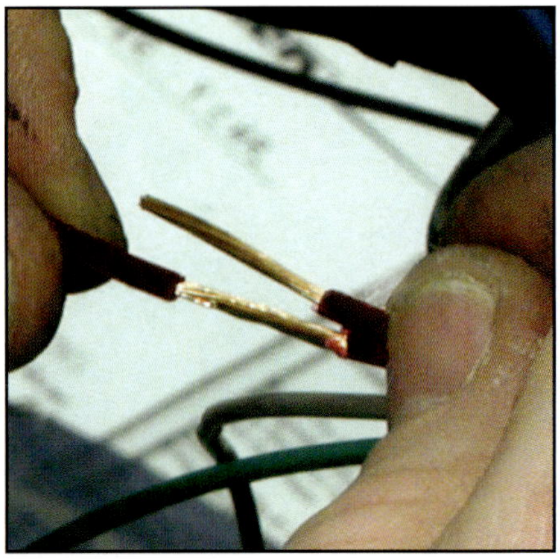

Minimizing the amount of wires in the engine bay was accomplished by melding multiple ground wires into one. Here, two wires are being combined with another wire in this process.

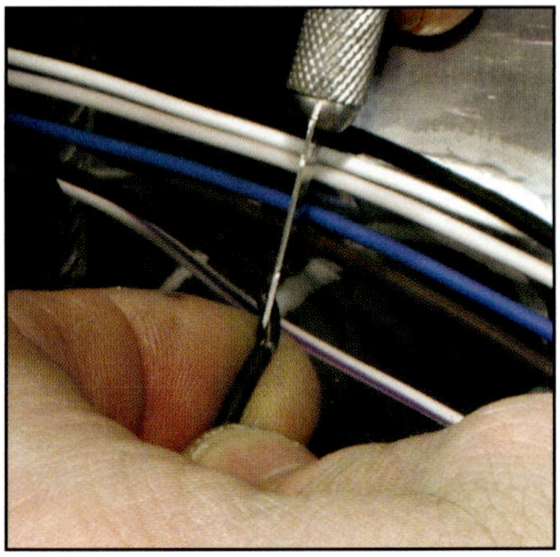

The first step is to cut a section of insulation off the main ground wire. Then, strip the end of the other wire(s) of insulation about ½ inch long. A 2 inch length of shrink tubing needs to be slid down the length of it on the 'open' end of the wire—so you can seal this section back up when the mating is complete.

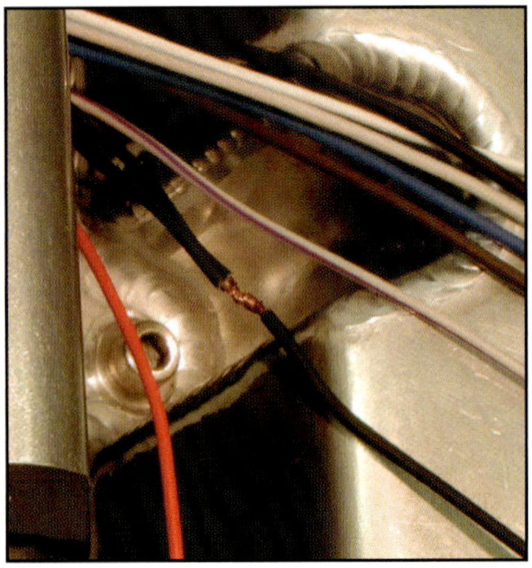

The other wires are then wrapped around the main wire at the stripped end, before being covered in molten solder to lock them together.

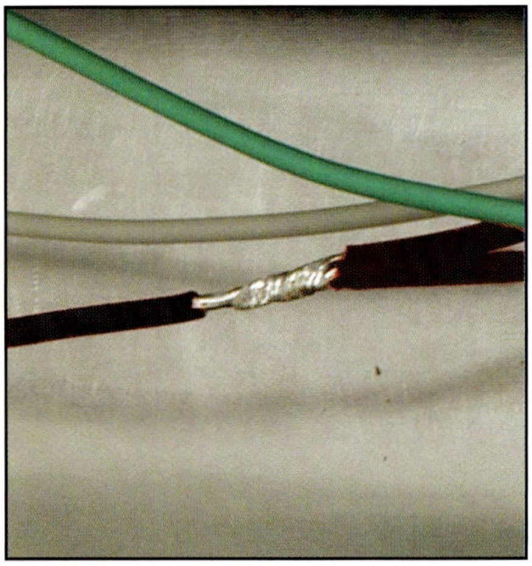

This is what the fused wires looked like before the shrink tubing was slid in from the left (the 'open' end) and heated up to shrink down on the wires.

Wiring up Military Spec Connector

The connector chosen as the bulkhead fitting for the engine harness uses wiring pins that are different from the Packard or Weatherpak pins, so an entirely different set of tools was required to install the pins…there are plenty of other tips here for anyone doing wiring.

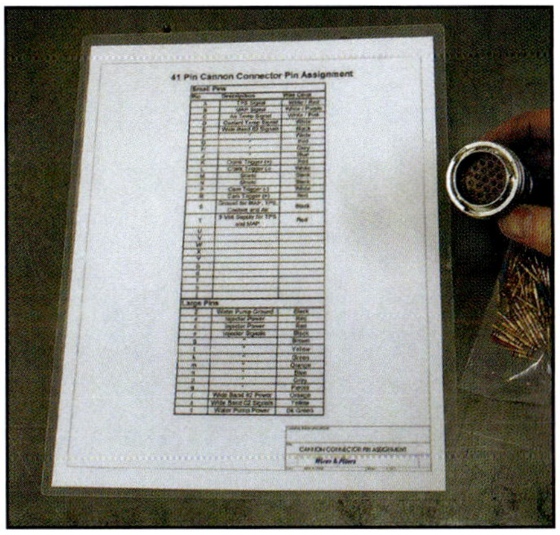

This is the proper way to start your wiring project: with a written plan for where the wiring pins are going to be located in the connector(s). The entire engine management wiring harness is being run through this 41-pin connector to speed up removing the engine.

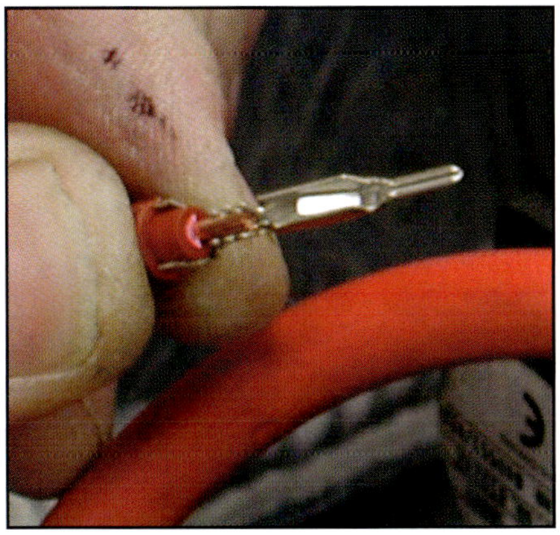

The male pins for this connector are similar in how they attach to the wire, but the ends are pointy…like a pin.

Like the previous wiring pins, a small grommet is slid on the wire, a ¼ inch of insulation is stripped off the wire, then the connector/grommet are crimped onto the wire using these special crimping pliers.

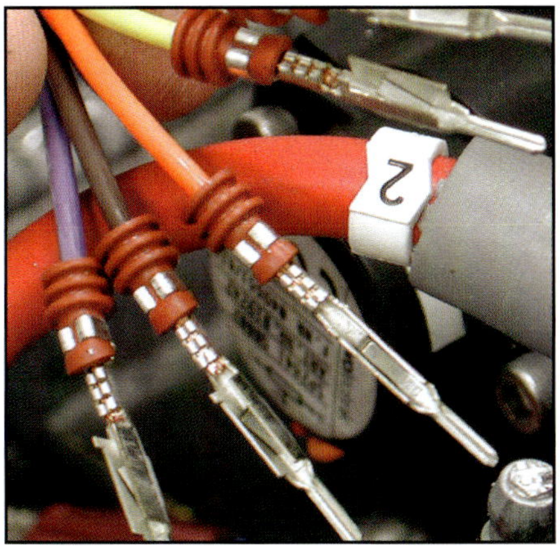

Here is what the male pins look like when properly crimped onto the wires. The female pins for the firewall-mounted connector are installed on their wires in a similar fashion.

Pros Trick: To minimize the stray RF energy created by the ignition coil discharges, the coil wires were wound in this helical form (arrow). This shape is held by installing shrink tubing over it.

Here, 3/16 inch diameter Raychem shrink tubing (that's the good stuff), has been slid over the two ignition wires and is being heated to set its shape.

This is what the ignition coil wire looked like after the connector was installed and all the shrink tubing was installed—looks 'right' doesn't it?

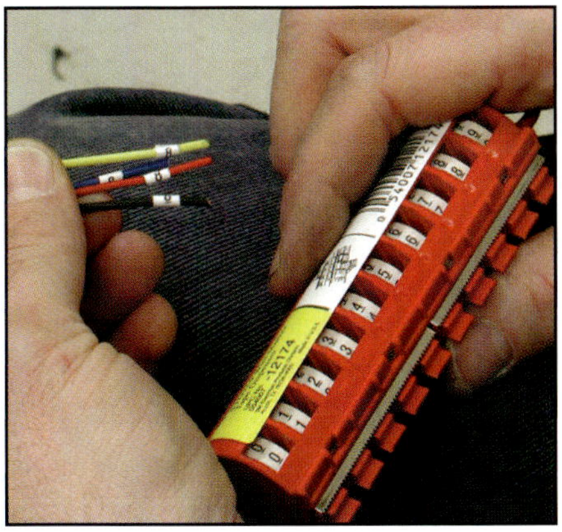

Don't let anyone tell you this if for amateurs: placing wiring numbering on various wires makes an insanely confusing job somewhat understandable. This cool dispenser spits out the numbers by depressing the little tab below the serrated metal piece.

This shot shows the bundle of wires coming together for the 41-pin connector. Notice how the wiring connector nut is already on the wiring bundle, along with a 1 ½ inch diameter piece of shrink tubing.

Back to the roadmap: the pros know this is the way to end up with a system that works the first time.

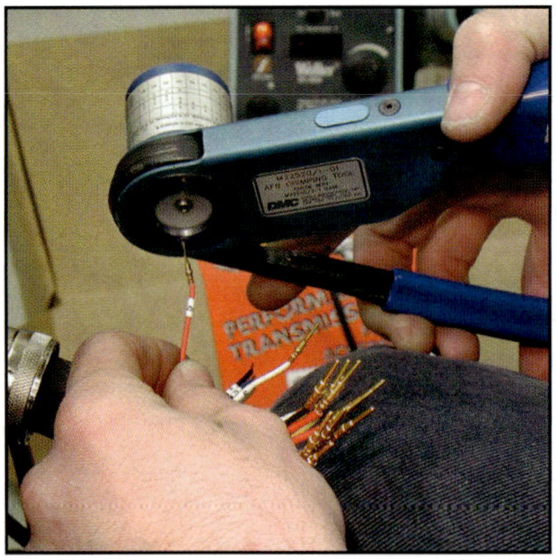

This crimper, from DMC (Daniels Manufacturing Corp, in Orlando, FL) is specially made to install these trick little wiring pins.

The bundle is now ready for install on the connector. Notice how there are subgroups of wiring ziptied together—this makes it easier to keep track of what wires are for what systems.

The pins are then installed in their pre-determined holes—this requires care to make sure they are not damaged during the install, but that they are fully seated in the connector. A delicate balance for sure!

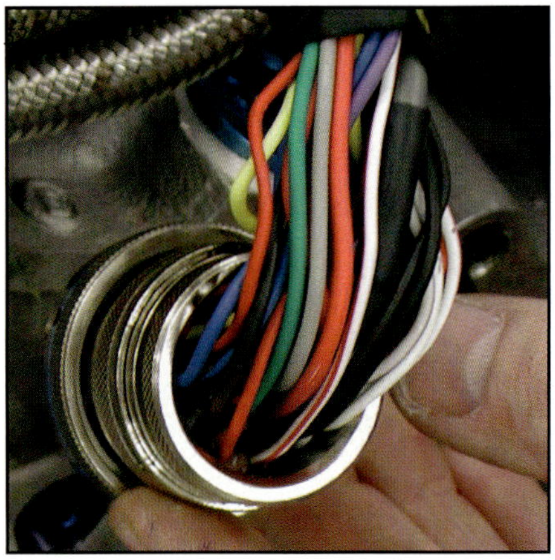

This gives you an idea how much wiring needs to be threaded through the connector nut—again, great care needs to be taken here to avoid peeling insulation off wires and causing a future failure point!

With the wires pulled through the connector nut and the nut threaded onto the connector, the shrink tubing was pulled forward over the connector and heated up to seal off this mating point. Notice how the rest of the bundle has been covered in large diameter shrink tubing.

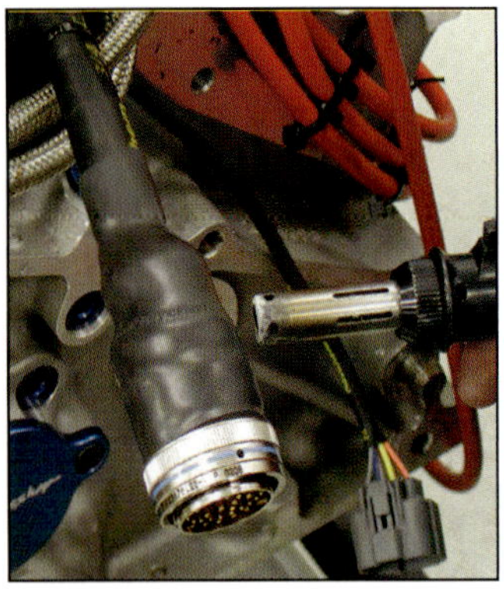

I am tough to impress, but can say this engine wiring harness came out better than I would have expected. It looks great, is practically invisible on the engine, installs/removes quickly and has worked perfectly all of these years.

Here is the inside of the engine compartment wiring harness connector. Notice that all the pins are not used…but most of them are.

Here is the female end of the 41-pin connector: it will be mounted to the firewall so it can act as a bulkhead passthrough for all the wiring. Note that the mounting plate has a 'flat' in it so the connector won't spin in the plate once mounted.

We installed the bulkhead passthrough connector about 4 inches below this small 't' so it could pass under the distributor. It is practically invisible in the engine compartment—my goal with the entire wiring system.

Sensor Tips

I don't like to look at sensors and wiring when the hood goes up, so I am always looking for places to stow them. As an example, I plumbed this aluminum block into the oil cooler lines so I could have an oil temperature sensor hidden out of the way.

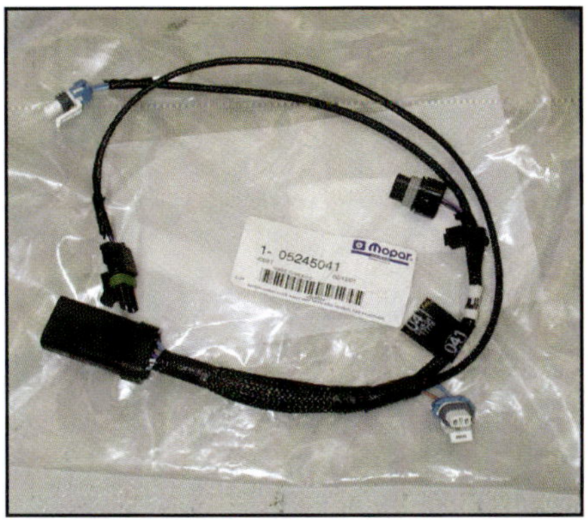

Since I have my T56 transmissions upgraded by D&D Performance in Wixom, MI with Viper componentry, getting the transmission speed sensor properly wired up requires using a Viper transmission harness—which I have taken a photo of here. This falls under the 'why make it if someone already offers it' mantra!

General Wiring Steps

Here is what it looks like when I am in the final steps of wiring up a vehicle. I can tell you having a wiring harness in my hands and figuring out the final few aspects before installing it is a GREAT feeling. Really.

This is the business end of the Painless Performance Products fuseblock. The really cool thing about this fuseblock is that it installs in the factory 'passthrough' opening in the firewall, but has almost twice as many circuits—that's the best of both worlds.

When doing wiring, give yourself plenty of flat space, either on the ground with packing blankets laid down, or on a large table. I prefer to work on the ground when wiring as I have the entire garage floor and it puts me close to my plastic tubs of parts, tools and pieces.

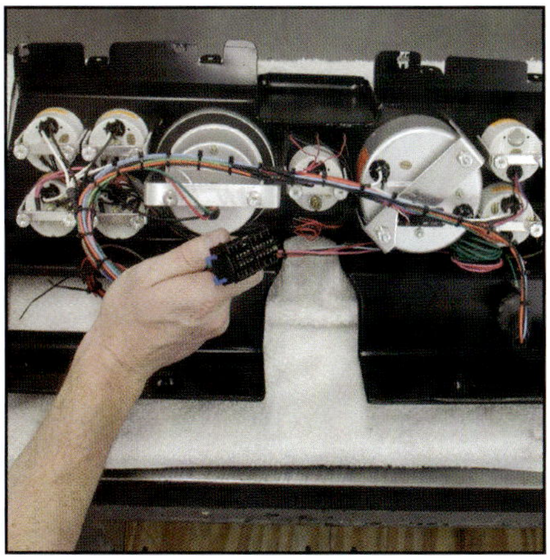

Yes, I build a harness specifically for the guage panel with its own multi-pin connector. This just makes it easier to pull the dash if I need to and also minimizes issues from arising later—and it looks cool as hell.

I find wiring to be an all-consuming activity. If you walk away for more than a day you forget what you were doing and where things were going—which is really, really frustrating. To avoid this, I clear my schedule and work day and night until this job is done.

I mount my Optima battery in the trunk area here (using a custom hardcoated DSE battery mount) routing 2 guage battery cable to the engine. On the Mule, I used an Aeromotive fuel pump that has multiple RPM settings—so I mounted the fuel pump controller near the battery and fuel pump (in the trunk).

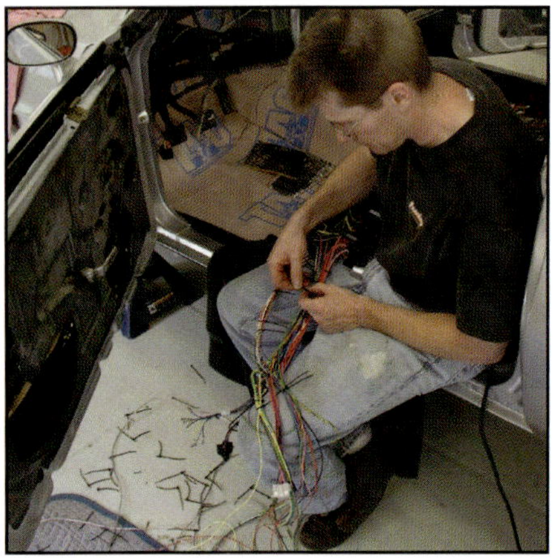

Once the wires are 'set' on the wiring harness, I spend considerable time getting all the routing 'right' on it. This is just cubic hours of trial and error with lots and lots of zipties being installed, the harness test fitted into the car, then removed and refined until it all fits perfectly. This is another really tough job mentally as there is no 'right' answer…just a lot of wrong answers!

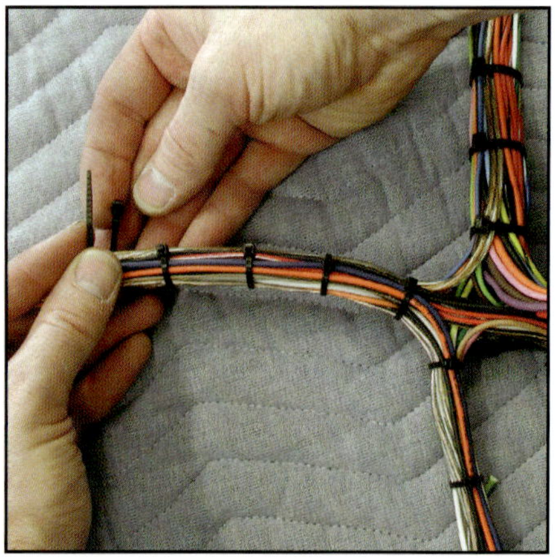

Once I have settled on the wiring routing, I lay the rough positioned wiring harness on the packing blankets on the floor and install zipties every 1 inch. I cut off all the excess ziptie lengths on each one. This is brutally mind numbing, but is also an exciting time as I know the end is near!

Then, I wrap the entire wiring harness in black loom tape, making sure to wrap around wires that split off the bundle in specific locations. This tape has no adhesive on it, so it is not electrical tape (even though it looks like it).

The trick with the loom tape is to overlap the tape onto the previous loop of tape about 50% while pulling the wires tight in the bundle.

At the other end of the harness, I wrap the loom tape around a few times on itself, then close off the loom tape with two or three loops of electrical tape (the additional adhesive on the electrical tape will hold it in place).

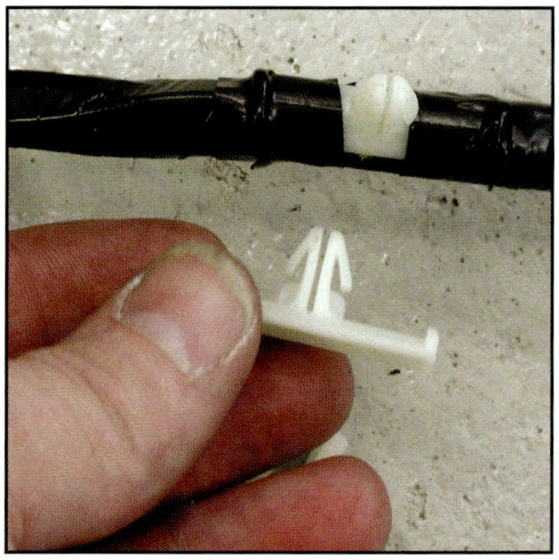

One of the tricks I have learned from taking apart late-model wiring harnesses is to wrap these wiring harness clips into the loom tape where they will meet up with holes in the body—this makes it easy to lock to location of the wiring harness down.

This is what the wiring harness looks like when wrapped in the loom tape. This was the rear wiring harness section—which is why it doesn't have any zipties on it (I put those on the main wiring harness).

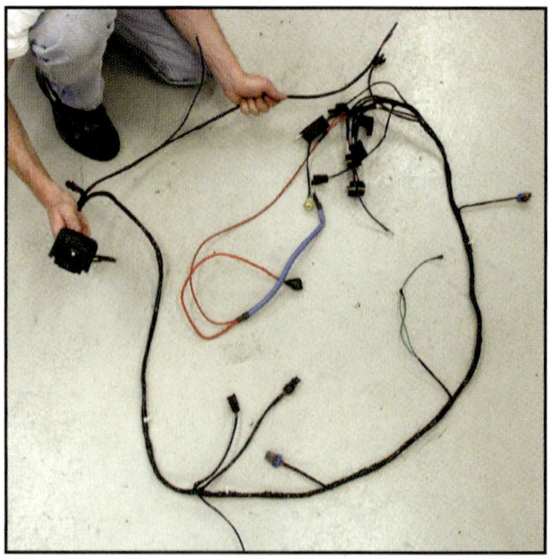

This is what my wiring harnesses look like when they are ready for final install. They look purposeful, only have what they need to get the job done and are as lightweight as I can make them (I am always looking to save weight!).

A Pro Touring Rear Seat

Shortened to fit

If you are building a Pro Touring car, mini-tubbing the rear wheelwells is usually a requirement (to stuff the 335 mm or bigger rear tires in the back of the car). Increasing the width of those wheelwells usually intrudes on the rear seat frame—and yes, a lot of my cars do have the stock rear seat in them. So, to get the seat frame to fit in the vehicle, the rear seat springs/frame needs to be modified to fit. This section shows how I achieve that, along with some tips for how to match the upholstery of performance front seats (like those available from Recaro or Sparco).

Getting the stock seat frame to fit like this around the four-link rear suspension, mini-tubbed rear wheelwells and rollcage requires a little modification! Here's how I usually do it…

The first step is to remove the stock seat covering—I have never reused the stock seat covers on my Pro Touring cars and this one is no different! Tear it off and keep it to give to the trim shop in case they want it as a pattern.

I start the seat teardown process by cutting the 'hog rings' off the perimeter of the seat frame with a set of side cutters. They can also be twisted off with a set of pliers, but I have bent the frame doing that. Wear safety glasses as these pieces do go flying…

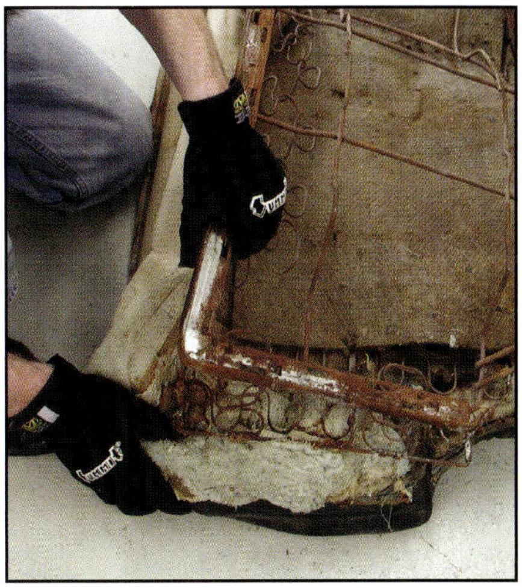

I then start peeling the cover, padding, jute and various fabric material off the seat. You will discover more hog rings as you peel, so be prepared to cut more off as you go.

This is what you need to get to: the bare seat frame & spring set. Check for broken springs. I have found the rear seats on these cars are usually barely used, so most are in perfect condition, but it is a good thing to check and fix now if there are any.

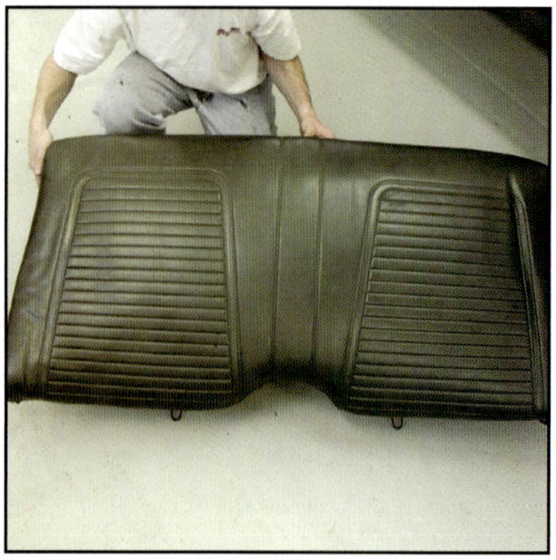

The rear seat back needs to be stripped, too, as it will 'crash' on the inner wheelwells that have been widened.

Here is how to twist the hog rings off: pinch down on them with a visegrip then twist. While the cutting method sprays hog ring pieces all over, I find that to be the better method as the chance of damaging the seat frame to be limited.

Here's another good tip: I usually work from one end of the seat to the other freeing the cover from the frame. This way, you can see your progress better.

While I usually do this type of work before the car is painted, sometimes that doesn't work out. When working around a painted car, it is always smart to protect as many surfaces as possible with foam wrap and tape—as is being done here on the rollcage bar.

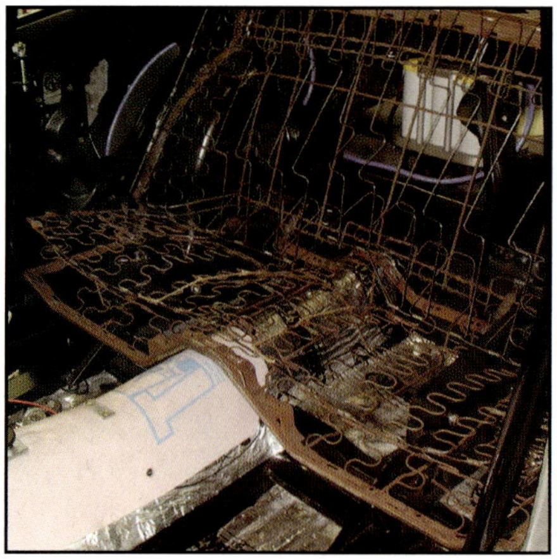

I then do a basic mockup of the rear seat frame back in the vehicle to see what is preventing it from sitting properly in the vehicle. I mark these areas for cutting and start that process of making it fit.

I like to measure twice and cut a little, then refit. In fact, I am VERY conservative when cutting something down to fit. I figure the time spent here is much less than what it would take to add material back on and make it look 'right'. I use a rotary saw to trim the frame.

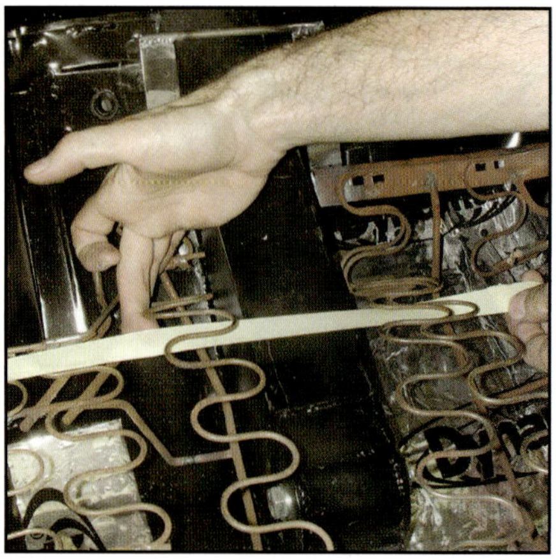

I refit the seat frame and use some masking tape to determine if the seat cover will clear the suspension load boxes (NOTE: these are my custom lower suspension link load boxes in the Mule…the DSE rear suspension lower link boxes are much smaller than these, so fitting the rear seat is much easier with them).

I then weld in a basic framework and refit the seat… repeating this process until it all fits properly. Notice I am using a magnetic block to hold the section wire in position.

Here are the pieces I needed to cut out of the stock seat back to get it to fit around the now wider inner wheelwells. This does not impact the appearance of the rear seat or how comfortable the seat is—it just puts less structure in the seat frame.

Sealing it Off
Whatever you are going to do with your Pro Touring car, I suggest you take the time at this point to seal off the trunk area from the interior with some sheetmetal over the seat back and speaker shelf area: this helps to seal off the trunk area from the interior. Sealing the interior makes the A/C work much better.

Water Papers
Another OE trick is to install 'water papers' over the openings for the side windows to minimize the chances of any water getting into the interior. Glue has been sprayed on this so jute can be stuck to it (for sound deadening).

Matching the Seat Fabric

I like to run performance seats in the front of my cars, but don't like the mismatch of fabric when the stock rear seat covers are used, so here is how I get everything to look good when it is completed.

The stock seat cover(s) were then cut up to be used as templates by my interior guy, Jerry Stewart of All American Auto Upholstery.

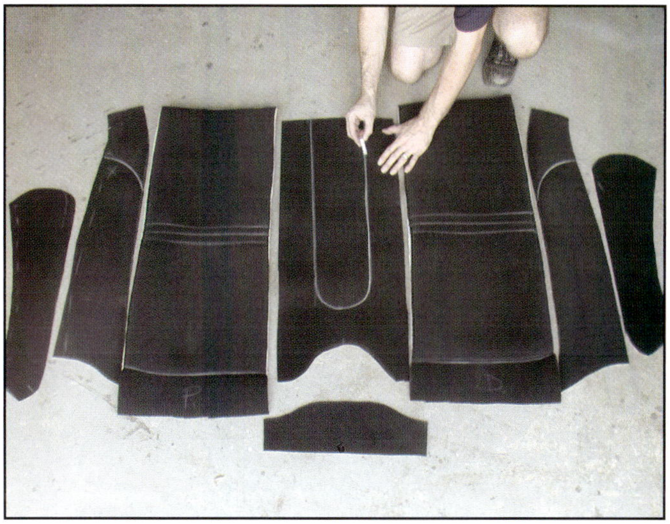

Using fabric purchased from Sparco (prepare yourself for the cost of this, but it is worth it in the look as far as I'm concerned), the pieces are cut to fit the soon to be recovered seat frames.

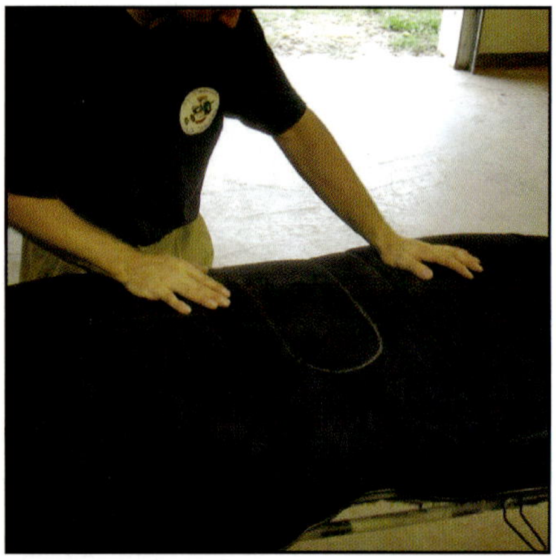

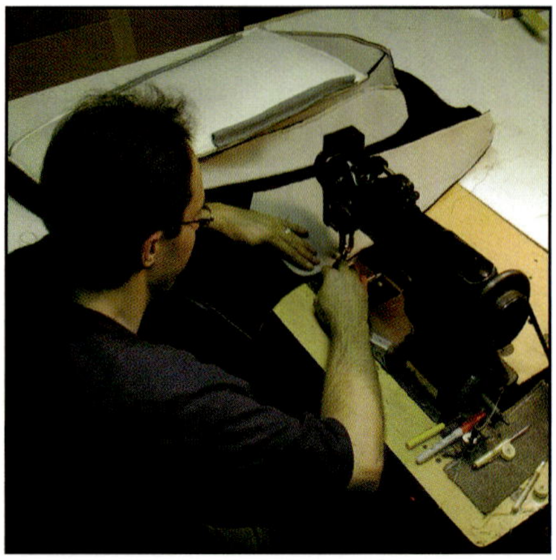

Just to clarify: I don't do interiors, but I like them nice. Jerry Stewart is just as particular as me, so we work well together. Here, he is sewing up the rear seat cover to replicate the front Sparco seats.

Mocking up the seat covers on the seat frames and getting the foam cut to 'form' around the seat covers requires skills that come from years of experience…but I can tell you it takes time and effort to get it right!

After extensive mocking up and reworking, the final shape is settled upon and the seat cover final sewn up.

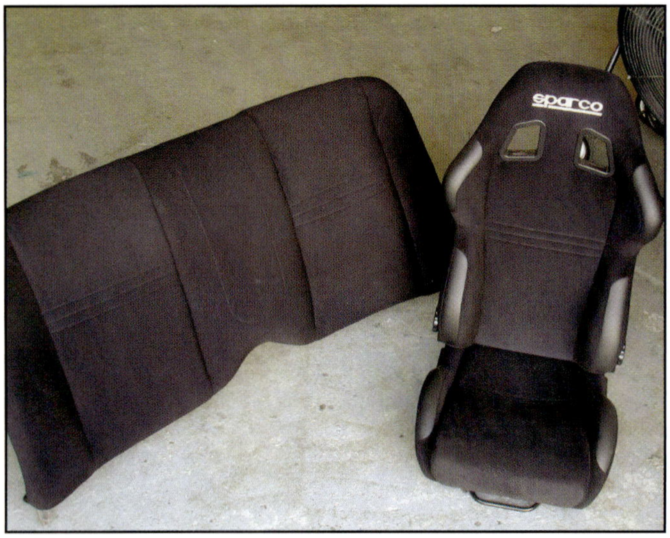

Here is a front seat with the now-matching rear seat back. Looks good, huh? It is even comfortable to sit in (for the highly cramped Camaro backseat). And yes, I do occasionally have people back there.

This is what the seat back and bottom ended up looking like once recovered.

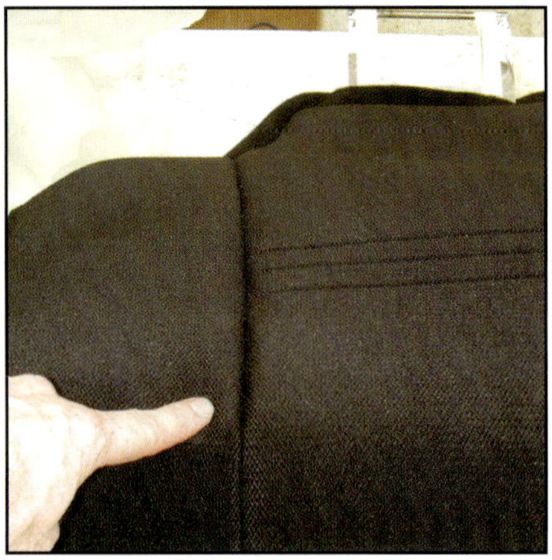

Jerry is experienced enough he knows that the rear seat back needs a 'spreader seam' in the cover to be able to flex when people sit on it—which is what he is pointing to here.

Interior Buildup

Some stock pieces, some aftermarket: light, reliable and simple

I know there are a million things you can do to hot rod interiors to make them 'cool': booming stereos, door poppers, custom interior panels, video screens, etc. While I like all those things, none of them make a car go any faster and probably won't help with making it more reliable, either, so I don't run them. It is just a personal preference. You will still need to consider the many things that are considered requirements by many folks—like side glass that goes up and down and trim pieces that actually look like they were meant for your vehicle. I spend a lot of time making sure those elements will work correctly--and will show you here what that prepwork looks like.

A little forewarning: there is a lot of detail work to the interior components (sure seems like there is a LOT of details when building a Pro Touring car, huh?). While this is still work, I like doing it because every time I walk in the shop at this point in a buildup, the car looks like it could drive out tomorrow…except for the missing side glass and some interior pieces. Whether it is true or not, I feel like I'm close to being done.

Deep down, I know the car still needs a lot of work—but I keep thinking about the car being done and that lets the time pass as I do the detail work that will make the car that much more fun to have on the road.

Do yourself a favor and tune up these interior pieces before getting the car on the road. They are some of the most often used, never heralded components & systems in your interior—my guess is you will be glad you spent the time to get them 'right' before putting the car on the road.

A lot of my vehicles have a similar looking interior: a mixture of mostly original equipment reproduction parts with some aftermarket and late model parts and pieces to improve functionality. This shot is of the Mule, a 1969 Camaro.

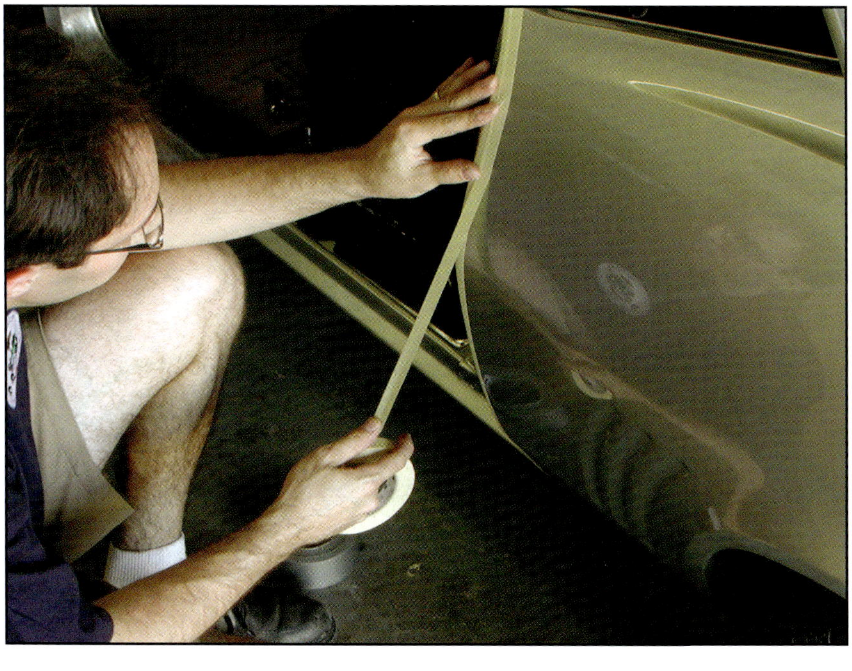

Interior Work Pre-Prep

OK, you went to the trouble of painting and plating everything so it looks nice, now the least you can do is cover the edges of the 'at risk' spots with tape and foam to minimize the chances of those beautiful surfaces being marred! Do it before starting out…

Front Seat Install

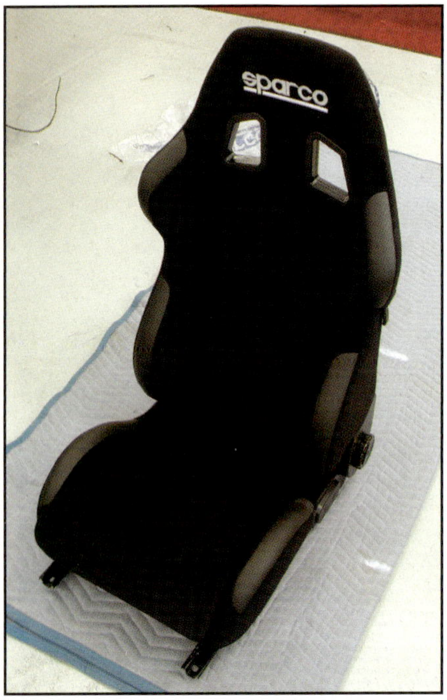

I like to sit in performance oriented seats that hold me tightly but are still comfortable for long drives—like these Sparco seats. But most of these types of seats do not 'bolt-in' to a musclecar. A set of custom adapter mounts usually needs to be created.

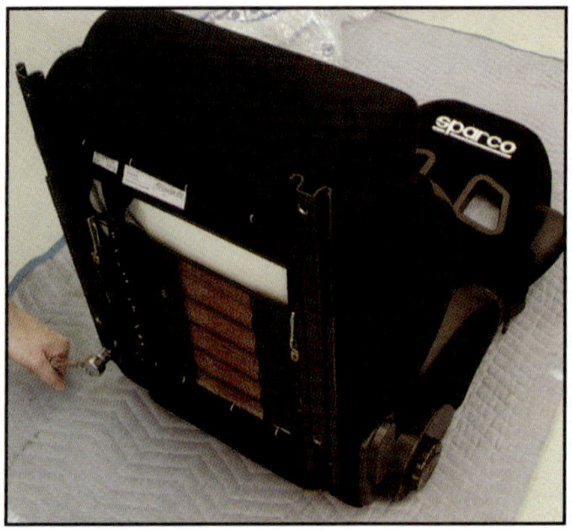

I always run seat adjuster tracks on my front seats—fixed seats are easier but very aggrevating in a 'street' car. Most of the time, I use a set of tracks from a seat manufacturer—on the Mule, I used adjuster tracks from Sparco.

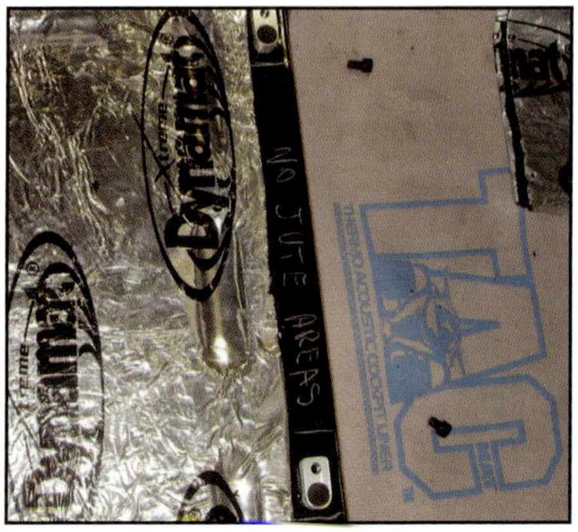

The seat track mounting holes were close to the spacing of the stock Camaro seat tracks…but different enough that these adapters (shown) were needed. I made them out of aluminum for weight savings, but could have made them out of steel if needed.

A countersunk Allen 5/16 inch bolt was used to mount the seat track adapter to the floor. This kept the added height to a minimum. This keeps my seats as low as possible in the car so I have good visibility out the windows and keeps my weight low.

Headliner Install

I run the repop Year One stock-type headliners in my Pro Touring cars. Until someone comes up with a lighter weight, more reliable and/or better looking setup (don't bet on it!), this is the best choice, in my opinion.

While I have professional interior guy (Jerry Stewart) install my Year One headliner, its good to know how the pros do this job. First, they lay out the new headliner, position the bows (always save you headliner bows in 'order'!) in the headliner and mark the front of it.

Jerry applies a light touch of Vaseline to all the headliner bow clips in the roof to maximize the chances of the bows sliding into the clips with no friction.

The headliner is then loaded into the car through the rear window opening and the bows installed in the center roof clips and holes on the sides. Once everything is confirmed in the correct location, the process of gluing the edges of the headliner to the roof starts. Note: this is easier with no glass in the car.

I don't know why, but the type of glue the interior guys use seems to be stronger than what you get in spray cans. That's another reason I have the pros do this step as a headliner coming loose is very annoying. Here, Jerry is starting out by gluing the rear of the headliner in place.

Next, the front of the headliner is glued down. The process involves spraying both the fabric and the gutter channel on the roof with glue, letting both surfaces 'tack' dry, then pulling the headliner tight before sticking the headliner to the sprayed metal.

One trick is that on the 'curved' surfaces, the headliner edges are cut into strips to eliminate any 'puckers' in the fabric. This is where experience really shows—Jerry cuts at the center of the curved areas…but getting the headliner to 'lay flat' requires a balance to how much you pull on the fabric and where you pull/cut. I leave this to the pros…

The other added headache to installing a headliner in a Pro Touring car is the rollcage tubing. The tubes make pulling the headliner up to the edges to get everything to lay 'flat' that much harder.

Fitting Trim

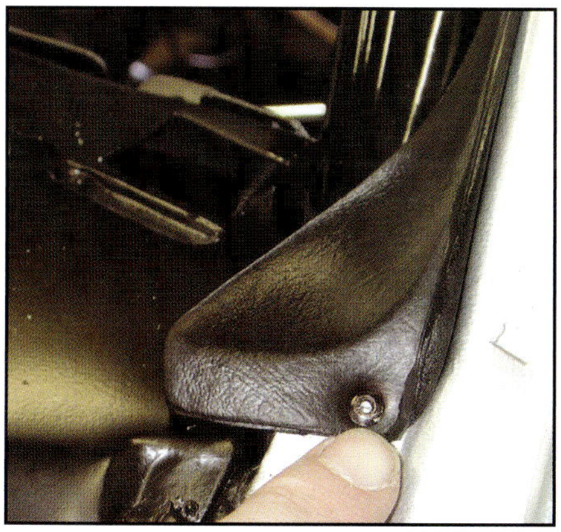

There are trim 'closeout' pieces that need to be installed all along the perimeter of the headliner to hide the edges. As is not unusual, the hole in the trim piece does not have a hole in the body that lines up. Stewart takes some time rechecking the positioning first, then drills a hole if he needs to.

Even though my cars are built to go as fast as possible, I do drive them on the street. Because of this, I keep the stock sun visors. It is difficult to package them around the rollcage tubes, but what I usually do is set the front hoop bar back enough to clear the visor…

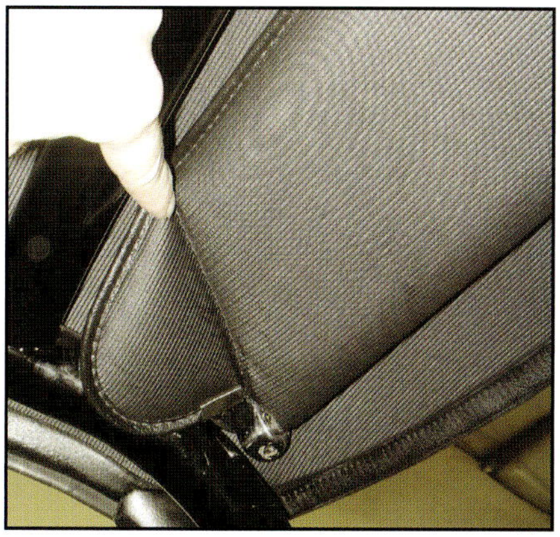

We sewed a 'break' into the visor so it would have a natural place to bend to clear the roll cage.

Rollcage Fitment Issues

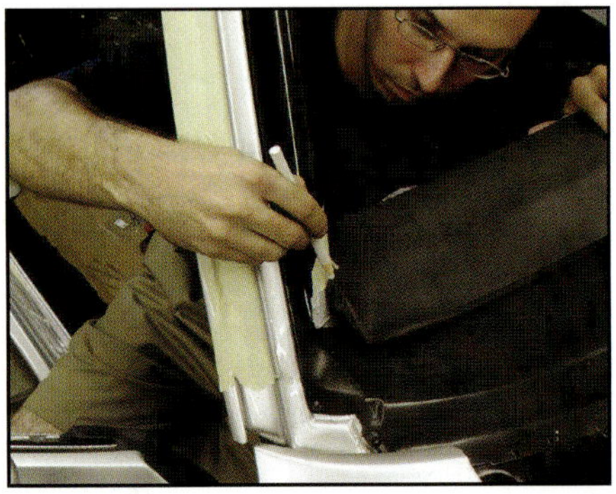

Another problem generated by the rollcage tubing is getting the interior trim to fit around it. This can be especially tough at the dashpad. The best method I have seen is to first test fit the interior piece (notice the tape protecting the rollcage bar from damage).

I recommend taking plenty of measurements to understand where the rollcage bar will 'pass through' the interior component. This will help you devise a plan to cut out the part of the interior trim piece that is hitting the rollcage...and also make any cuts needed to allow the interior component to be installed around the rollcage (usually the tougher part!).

The best cuts seem to come from a pneumatic saber saw. A good tip is to position the saw blade so it makes the cut at the same angle the rollcage tube is going to pass through the interior part. This positions the tube as tightly against the interior part as possible.

Side Glass and Door Tuneup & Install

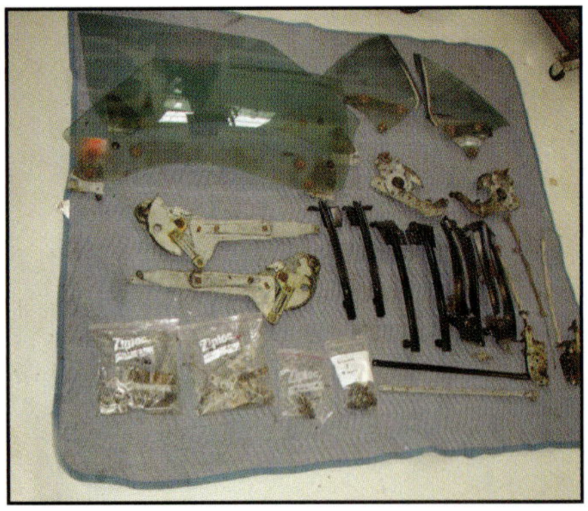

I like my side glass to go up and down effortlessly—and look good, also. To make this happen, I do a lot of cleaning, lubing and finessing. None if it is especially difficult, just time consuming. I suggest you spend the time on this now as you won't want to do it later!

TIP: When disassembling a car, I suggest using the tried-and-true method of bagging components and fasteners in ziptop freezer baggies (the freezer bags are thicker plastic) and marking the bags with notes on their contents with Sharpie pen. It's also good to date them so you remember when you removed them.

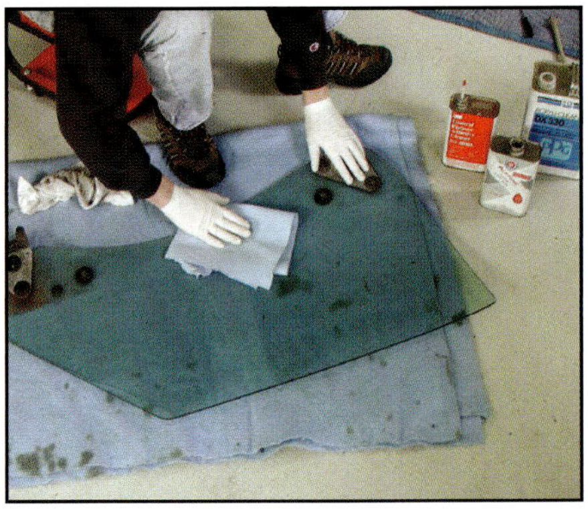

With the stock side glass parts on packing blankets, I clean everything with solvent to remove the gunk and goop that prevent the parts from moving smoothly and looking good. I check the actuation on them—usually the driver's door parts are worn out but everything else is in good condition. I replace worn parts with Year One pieces.

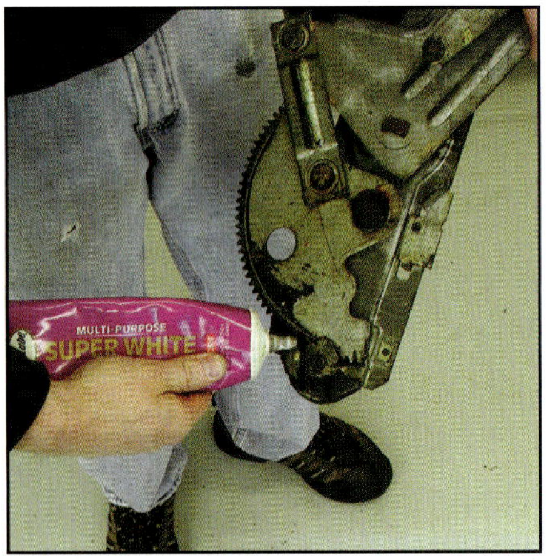

I then start the side glass install by applying a thick bead of multi-purpose Super White grease to the gear teeth of the window risers—and yes I use the stock hand crank mechanisms on my Pro Touring cars. They are light, easy to use and I have never had a problem with one.

I then install the window risers. A Tip: if you are not familiar with the car you are working on, I suggest seeking out an owner's manual, taking pictures of an existing car or taking pictures of the disassembly (I do this a lot) process to ease the install process.

I install the glass run channels in the doors and loosely bolt them into place.

I then load the side glass into the door—sliding it down the glass run channels and bolting it up to the window riser anchors. Note for later: Once all the door and window seals are installed I will tune the location of the glass run channels and the window riser mechanism to get the glass working smoothly and sealing tight.

The rear ¼ window side glass then needs to be installed. I do the same process of applying a thick bead of white grease to the window riser mechanism (after it has been thoroughly cleaned of grease/gunk/etc).

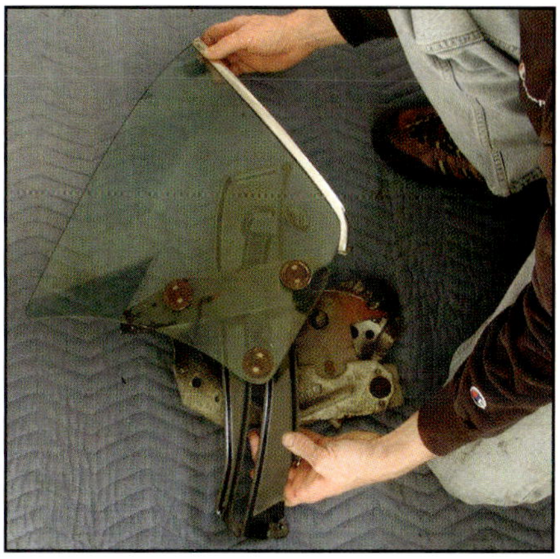

The next step is to assemble the window riser mechanism and glass into one unit and carefully load it into the window opening in the body. As a tip, I lay towels and rags over the 'show surfaces' when installing anything to minimize the chances of chipping paint.

Once the glass/window riser mechanism in lowered into position, I hand start the mounting bolts (fingers pointing to them) and lightly torque them—the position of this assembly will also be tuned once all the seals and channels have been installed.

Preparing to Install Door and Glass Seals

The next step is to install all the door and glass seals in the body channels. I always, ALWAYS, get new seals from Year One as any old rubber seal is junk. Also, make sure you have a new tube of 3M Super Weatherstrip Adhesive on hand to glue the seals into place.

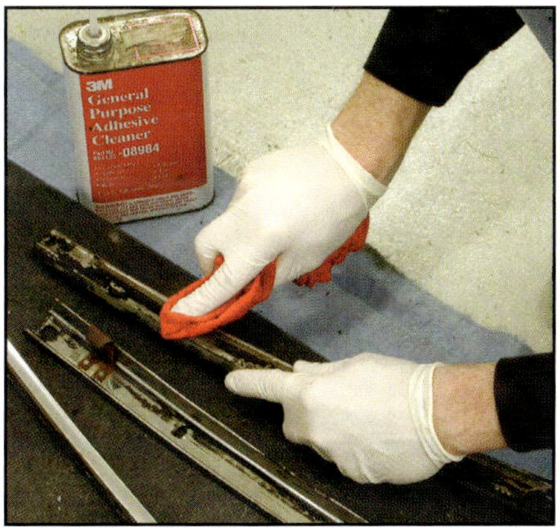

As with everything else, the first step is to clean the decades of gunk/goo off parts. Probably the best product for this is 3M General Purpose Adhesive Cleaner or Acetone. Use this product in a well ventilated area with gloves on your hands and plenty of fresh rags.

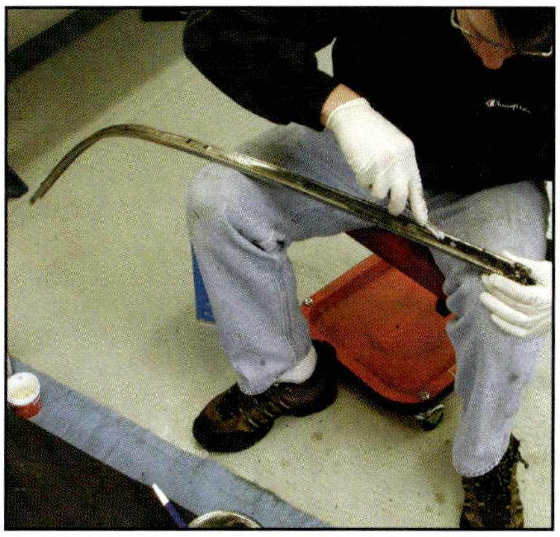

Sometimes, the scounge in the trim pieces requires more than a rag to remove. In those cases I will carefully use a razor blade, scraper or if it is in an area that is not going to be visible once the new seal is installed, a small metal brush—all soaked in solvent.

Preparing to Install Door and Glass Seals (continued)

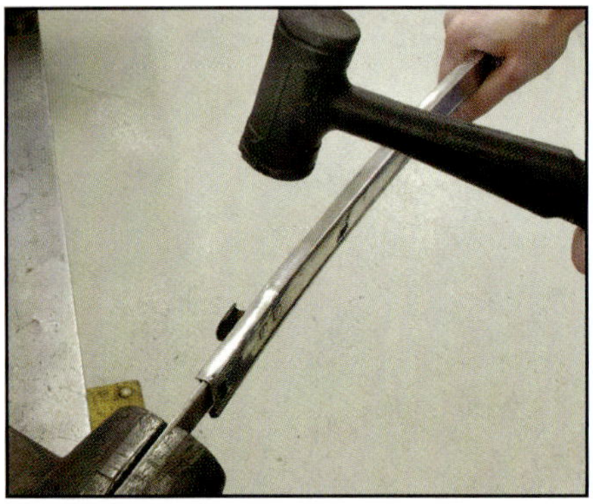

This is not for the fainthearted: if there are dents in my stainless trim/seal channels, I will tune them up by placing an 1/8 inch thick mild steel strap in my vise, sliding the channel over the bar and 'flattening' out the dent or ding. This takes patience but is not necessarily hard. There are also stainless trim shops that will fix these...

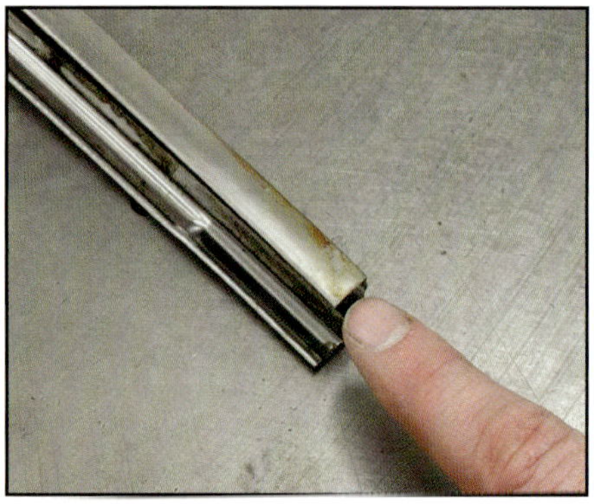

Sometimes, a stainless steel trim piece will have a little rust or discoloration on it like this small area near its lower edge. This can be fixed with some careful polishing work.

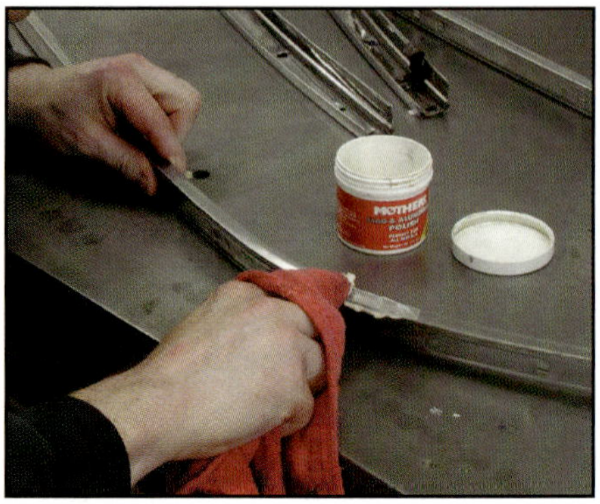

I hand polish the 'show surfaces' with Mothers polish until it looks new again—which is one great thing about stainless trim pieces, they can be tuned up in your garage to look great.

Door Seal Install

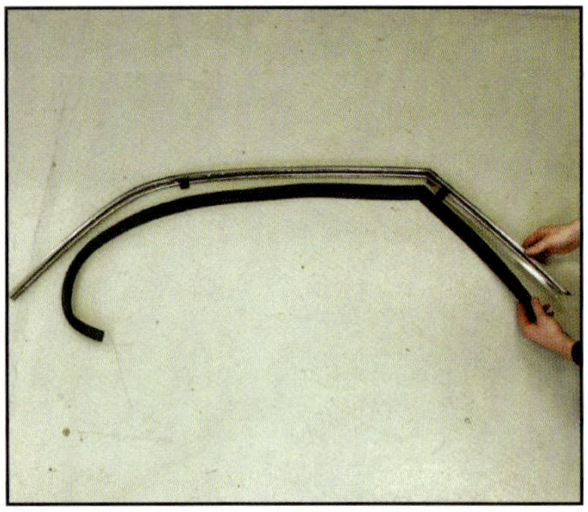

I then mockup the seals in the seal channel/trim pieces to make sure I have everything situated correctly—setting it all aside until I need to install them on the car. Mistaken identity is one of the toughest parts about installing seals: they come with limited instructions and all look similar. Take your time here as you can easily install a seal in the wrong location if you are not careful!

I then install two-sided, adhesive coated foam on the backside of the stainless trim pieces in preparation for installing them on the car. This foam will seal the trim to the body so water will not run underneath them into the car in a car wash or rainstorm.

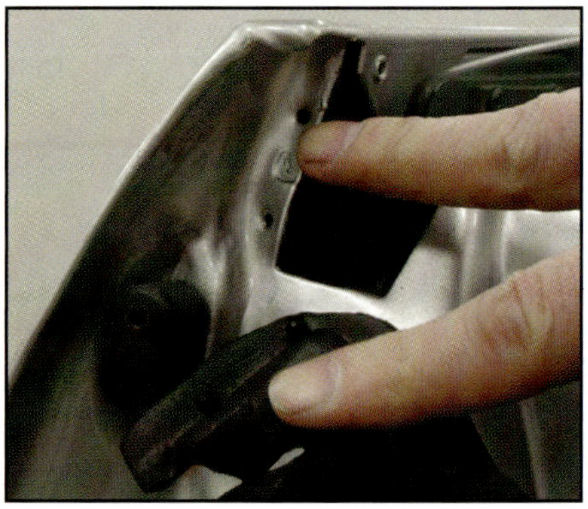

The door seals then need to be installed. This should be an easy job but remember, these cars were built in the '60s—there is a lot of variation in the builds. This upper rear corner door seal has three screw holes that should line up and go in...but they didn't in this case.

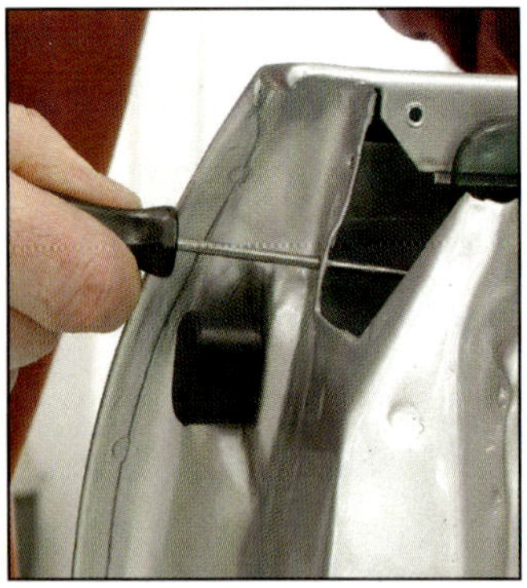

Sometimes, a hole's location can be 'tuned' slightly by installing an ice pick into the hole and slightly pivoting the pick in the direction you want to 'move' the hole. This is only for very slight adjustments—but does work.

The nuclear option to fixing a bad hole location is to weld up the existing hole, grind the weld flat and redrill the hole (shown) in the position the seal wants to be located in. This is a pain, but I want the seals to work/look right, so if it is what needs to be done, I do it.

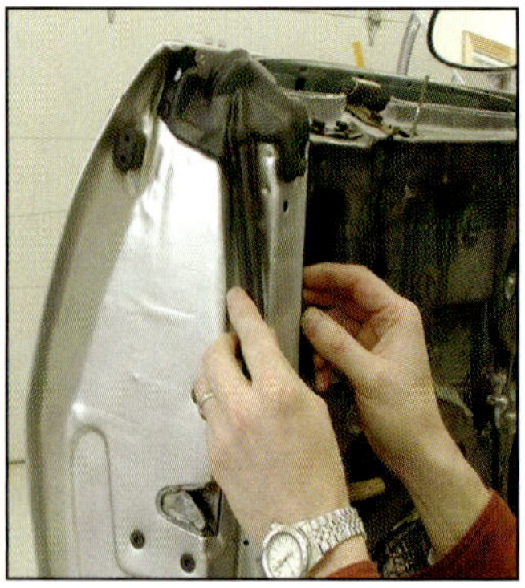

Once the corner piece for the door seal is screwed into position, I mockup the rest of the seal on the door by taping in onto the door with masking tape, checking the position of the clips to the door holes as I go and making sure it fits the door correctly.

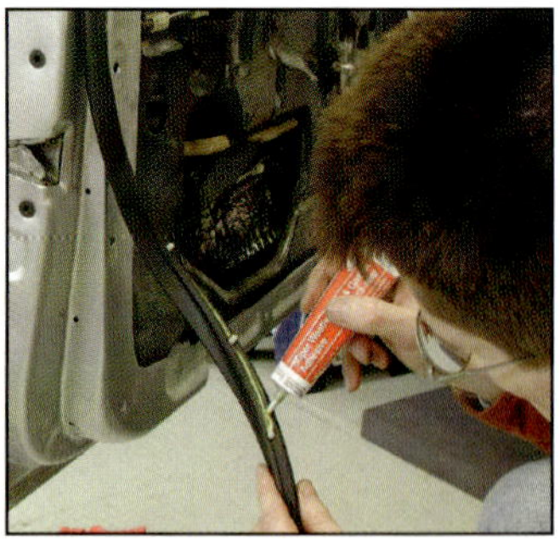

Once I am confident the seal will fit right, I pull the tape off and get out the adhesive. I apply a 1/8 inch bead of adhesive down the center of the door seal between the clips, then install that section, then apply adhesive to another section of the seal and push those clips in. I keep doing this until the entire seal is installed then use masking tape to hold the seal tight against the sheetmetal for 24 hours to make sure the adhesive has bonded.

Setting Door Locations

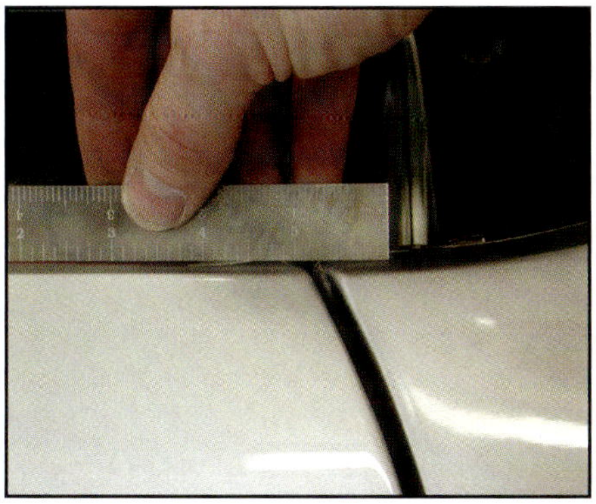

The location of the doors are then checked to make sure they are flush with the rest of the body. These are usually tuned by the painter, but I always check them to make sure…

If you do need to make a move, these hinge bolts will need to be loosened to do so. I am always very conservative making a change here as a little move at these hinges makes a lot of difference at the other end of the door!

Setting Side Glass Sealing

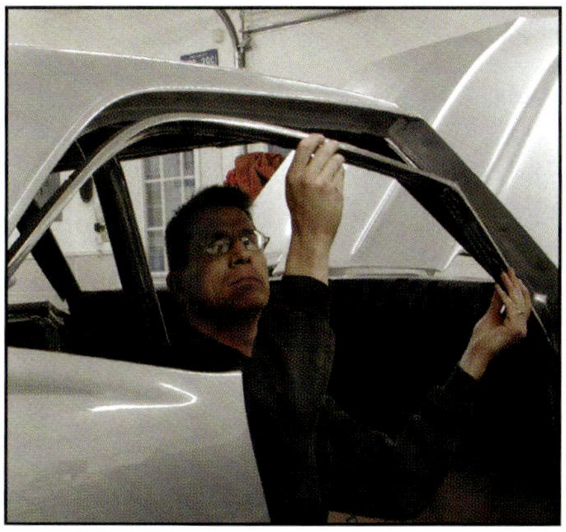

I then install the door seal trim/channels and door seals that I cleaned and prepped previously. If these came off the car originally, the screw holes should line up, but if they are repop parts, you will probably need to open up or outright drill some holes to get everything to go together.

I then push the door seal into the channel with a plastic interior tool starting from the front and working my way to the rear of the vehicle. These plastic tools are great because they do not damage the rubber parts.

The next step is to roll up the side glass to see if the glass comes in contact with the seal. I first just take a quick look at the glass to see if I'm even close. If the entire glass is not coming in contact with the door seal, I readjust the position of the window and door seal trim and then close the door and check it again.

Once I have the door glass to where it is touching most of the door seal, I get up close to the mating point with a flashlight to see where I need to 'tune' the location of the seal.

To seal out wind, water and dust, you need the glass to have a little tension on the seal. An easy way to check this is to place a dollar between the glass and seal, close the door on the dollar and pull the dollar out—it should 'drag' a little as you pull it out.

When I find a point where the seal and glass don't have enough tension, I peel the seal out of the trim channel, loosen the screw closest to that point and lightly pry the channel outboard while tightening the screw to lock it in place. I then repeat the dollar bill test/channel adjustment until there is even tension over the entire door seal.

Setting Side Glass Sealing (continued)

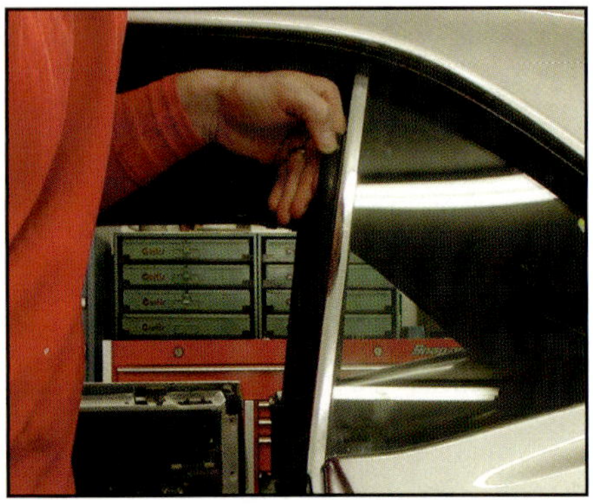

I then get the 'B-pillar' seal wet and slide it up onto the rear quarterglass channel and repeat the tuning of the seal-to-glass mating (dollar bill test, seal position tuning, repeat) until it all is 'right'.

By the way, the way to tune the seal tension on the B-pillar is with 'stuffers'—which are little pieces of foam stuffed between the sealing lip and the mounting lip to increase the tension and/or height. I am pointing to the two areas where I installed stuffers.

Installing Door Felts

The 'felts' are the large seals that mount on the top edge of the door on either side of the rollup glass. They are called 'felts' because the surface that touches the glass is actually felt-like to allow the glass to rub up against it without dragging.

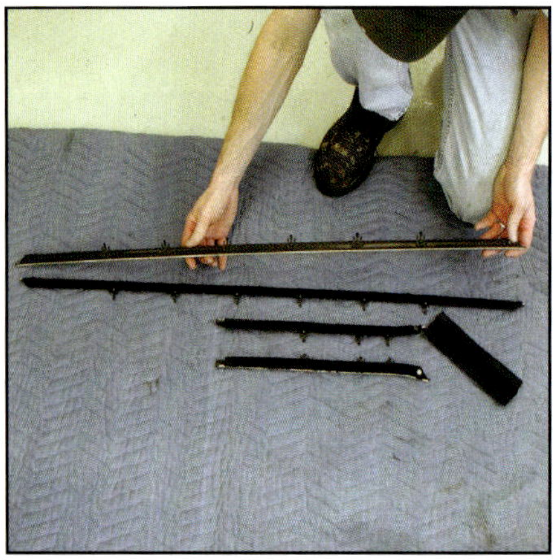

These felts are from Year One. They install easily and work great. I do usually need to spend time making them look 'right' on the car—because of 'build variation' again. All the doors are different lengths and the trim needs to be modified to fit correctly.

These clips are what holds the felts in place over the length of them, with a screw at the end to lock the position of the felt.

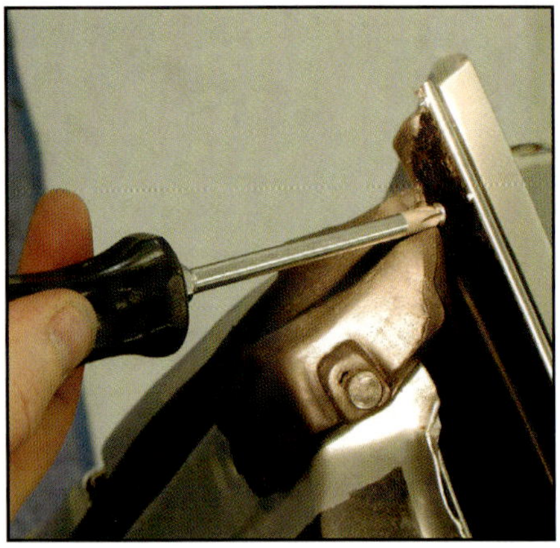

I do my felt install process by positioning the glass in its lowest position, setting the felts in the door glass opening, lining up the clips with their holes and then pushing the clips into their holes. The screws are then installed hand tight—so I can check the fitment.

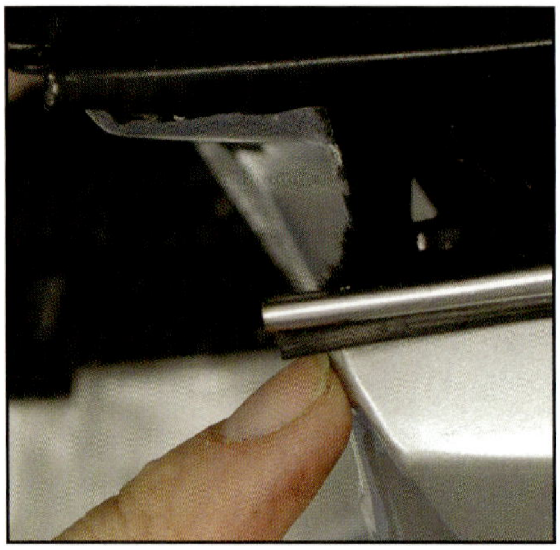

Again, this mismatch in the length of the felt trim is not unusual—and nothing to knock the reproduction part on either because these doors vary dramatically in length. I will now show you how I tune the length of these so they fit 'tight' to the door end. Place a mark on the felt where the door ends on it.

I remove the felt from the door and position it securely on a workbench (being careful not to damage the 'show surfaces'—both rubber and shiny metal), then get out a rotary saw to cut the metal trim.

Installing Door Felts (continued)

Once the metal is removed, I trim the rubber seal with a set of side cutters to match the new length of the felt trim.

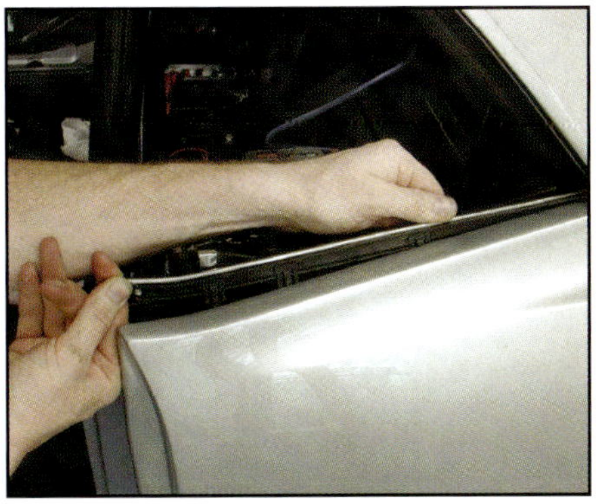

I then repeat the felt install process on the rear quarter windows until all four window opening felts fit tight and look great.

Door Panel Tuneup

Most of my cars use reproduction door panels for a few key reasons. I like the way they look, they do the job of sealing off the door with minimal weight, they fit with minimal work and are reasonably priced. They do require some work to install—which I will show you here how I do.

I start out by installing the various trim clips that are situated around the sides and bottom of the door panel. This usually involves cutting out the 'water paper' plastic that covers the backside of the door panel—being careful not to cut through the 'show' material on the other side of the doorpanel.

Here is what I like the clips to look like in the door panel before I install it on the door.

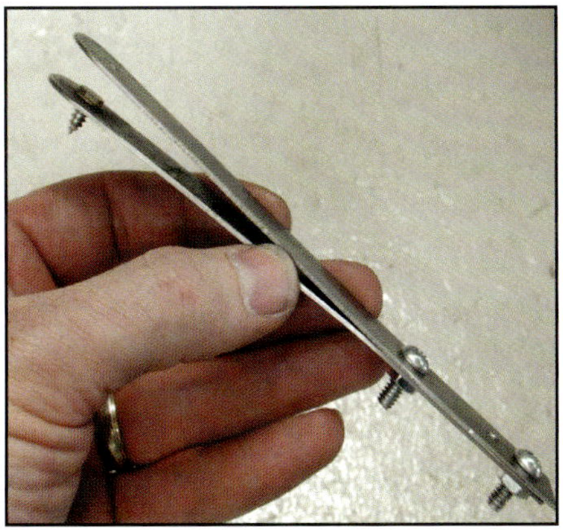

You know what this is? It is a Trim Screw Hole Locater. I built the one in this photo, but they are now available at www.techafx.com. This tool eliminates the chance of punching a trim screw hole in the wrong place—a BRUTAL mistake when final installing trim.

Here is how the Trim Screw Hole Locator works: you hang the door panel on the upper edge of the door and line up the clips with the openings in the doorskins. I then slide the TechAFX Trim Screw Hole Locator into position, setting the locator screw into the hole in the door. Next, I push a Pick tool to through the hole in the Trim Screw Hole Locator and the door panel show surface material and backing board. What used to be a terrifying job now just goes 1-2-3-Done.

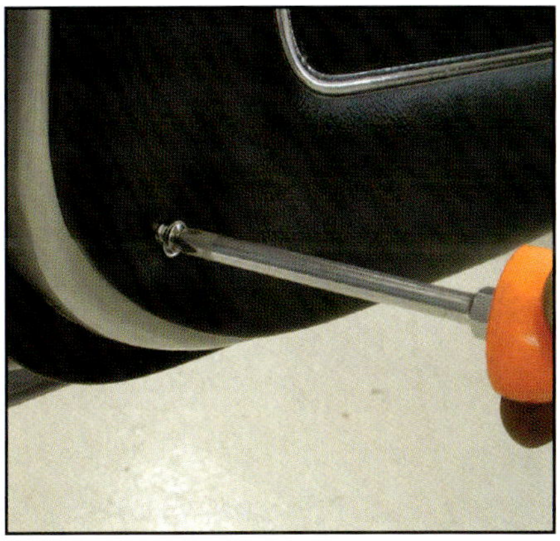

From there, I just install the screws hand tight and start checking positioning of the door panel on the doors. I usually find at least one area where the door panel is longer than it should be...again, just 'build variation' from the good ol' 60s. In these cases, I note the area that needs trimming and remove the trim panel.

Door Panel Tuneup (continued)

If it is needed, I start the trimming process by peeling the door panel vinyl covering off the area that needs to be trimmed.

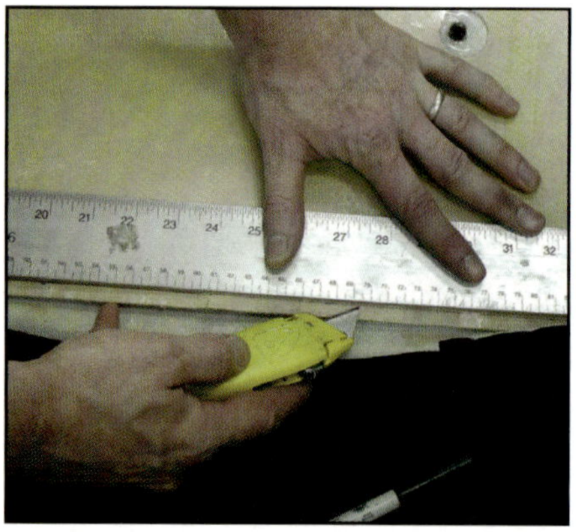

After positioning a few small screwdrivers between the vinyl covering and the backing board to create space, I carefully cut through ONLY the backing board of the door panel. This is definitely a time where I measure twice, cut conservatively, test fit and repeat until done.

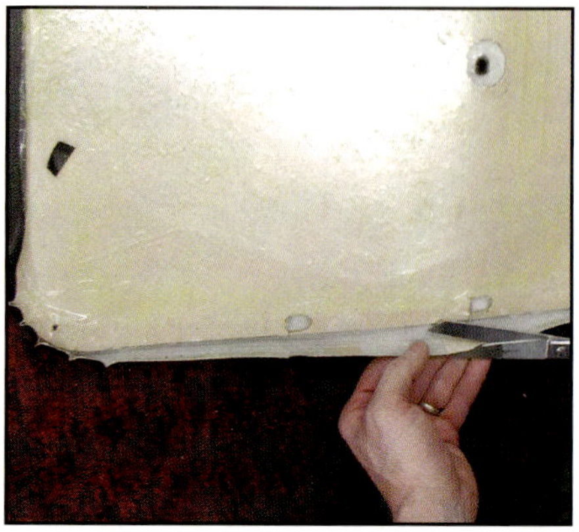

I trim the length of the vinyl covering just a little less than what I cut off the backing board in preparation for gluing it to the backing board.

To glue the vinyl back on, I lay the section of the door panel I need to glue the vinyl back on over a piece of cardboard and spray both the back of the vinyl and the backing board with upholstery adhesive. I wait until this glue has 'tack' dried: which means it is tacky when you touch it.

With the glue tacky, I carefully and lightly pull the vinyl tight around the edge of the backing board (to avoid any puckers or wrinkles in the material) and roll it onto the glue-coated backing board. Pushing the vinyl into the backing board to maximize the adhesion.

Door Handle Opening

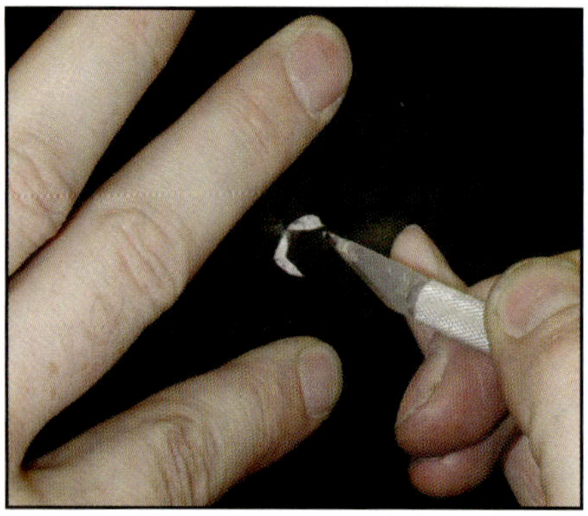

Most reproduction door panels do not come with the hole in the vinyl for the window crank—so this needs to be cut out. The backing board has a big hole in it, but the hole in the vinyl needs to be small. I do this cut by pushing the door panel vinyl into the window crank gear (with the door panel mock installed) to cause an impression in the vinyl. I then cut a hole on the inside edges of the impression with an Exacto knife.

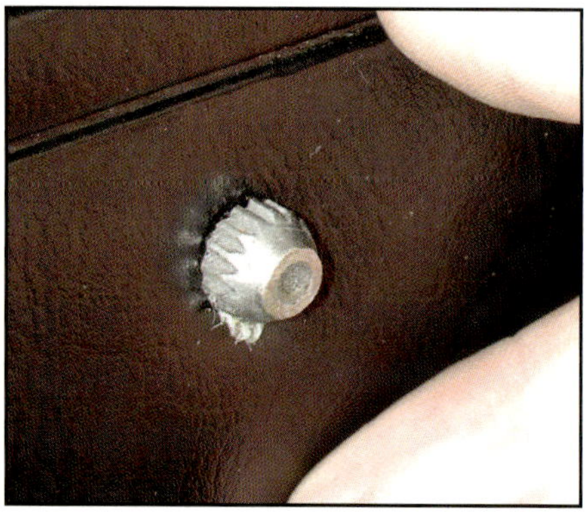

This is what the hole in the vinyl covering on the door panel should look like for the window crank pass-through. The tighter the better as this prevents any air from coming through this opening (annoying you) and also looks good.

Floor Pass-Through Install

Reinstalling the floor passthroughs is usually a no-brainer. Unless you have intruded on them with subframe connectors and rollcage tubing! (which my cars usually do). Here's how I get them back in the car the right way.

The GM factories in the '60s used body sealer or liquid butyl to seal these pass-throughs to the floor. I use these strips of butyl looped around on the mating surface of the pass-through. You can buy these butyl strips from Year One.

On the pass-through openings that are now filled partially with the Detroit Speed Subframe Connectors, I cut the pass-through the appropriate amount (measure twice, cut conservatively, test fit, repeat until right) in a band saw to minimize mutilating the thin sheetmetal part.

After applying copious amounts of butyl to the backside of the pass-through, I reinstall it in the stock location (but now cut down considerably).

Bumper Details

Getting the bumpers to fit 'right'

When I get to the point of having just a few pieces to bolt on a car before it is 'roadready', stopping to finesse those pieces in place so they look and work great takes much more discipline—because I just want to go drive the car. With that in mind, I know the bolt-on bumpers of early musclecars are a thing of beauty when finessed to fit properly on a vehicle. But when they don't fit right, they can be a powerful distraction on a good looking car.

Over the years, I have learned it all comes down to tuning the location of the mounts, and sometimes the bumper itself, to get it on the car correctly. There is no doubt that straight bumpers, whether they are chromed, blacked out or painted body color (which I have done all of), are a requirement to having a winning Pro Touring car (in looks).

Sitting Straight

The hardest aspect of installing early musclecar bumpers is getting them to sit 'flat' and 'square' on the body. On just about every car I have built, I have had to do some tweaking and tuning to the bumper shape and bracket mounting locations—and that includes both original equipment and reproduction bumpers.

The most common modifications are to the bracket hole openings to get the bumper to sit properly on the body. I think the majority of this mismatch is driven by the build variation in these vehicles, built over 40 years ago, and then probably added to by the abuse they experienced over the years.

The bumpers are somewhat flexible (except for the formed end shapes which are pretty well locked in position), so most of this process does not require serious force to change the bumper shape or placement, but it does take a lot of on-off-rework-on-remeasure-off-rework…

The look of properly fitting bumpers is critical to finishing up the 'look' of a real Pro Touring car. Like most of my cars, the Mule had a chromed metal bumper in the rear.

Endura Bumper Tuneup

I have built two cars with original Endura front bumpers and can say they are some of the most difficult to rework of the bumpers built in the '60s. The Endura bumper was an option on the '69 Camaro and a few other early musclecars. They are solid plastic bumpers with a metal frame molded into it…a precursor to the molded bumpers we now have today on all modern cars. Not too many of them were sold, so they are rare, and, as I found out, are rather difficult to work with.

Some of the reasons I would rather not work with the original Endura bumpers include the fact that an original Endura bumper cost well over $1000.00—and that's if you can find one! Also, they are stiffer than the metal bumpers, so tuning up their fitment on a vehicle is tougher. While I do like the way they look when finally fitting properly on a car, it is just waaaayyyy too stressful to get them sitting 'right' on a car. And being a weight freak doesn't help here as they weigh about 35 pounds—or considerably more than a steel front bumper (which weighs 15 pounds).

Getting the bumper gaps even to the body over the entire length of the bumper is a struggle. But it is worth it in the look of the vehicle. Also, having a bumper end that hangs at a weird angle is very distracting. Fixing this is very difficult, but doable. This type of fix needs to be done before you chrome the bumper. To do it, I pie-cut the bumper in the middle of the bend, change the bend angle to even out the bumper-to-body gap, then reweld/grind flat the cut area before having the bumper chromed. Obviously, if I can avoid this type of rework, I do!

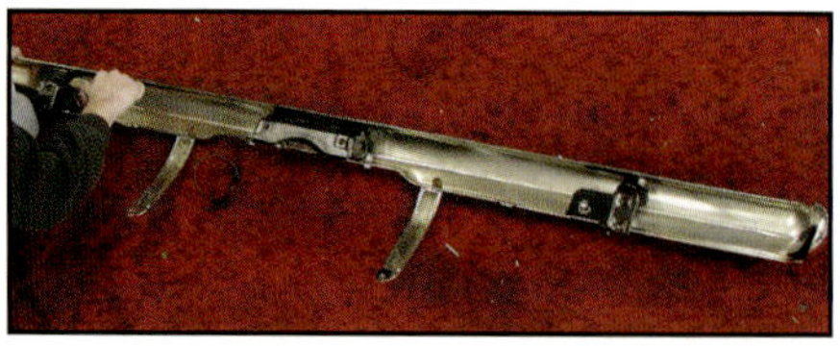

Getting a metal bumper to fit properly on a body requires tuning the bracket locations and possibly the bumper shape itself. I do a lot of this work when the body is in primer, but this process seems to always require additional finesse once the car is painted and in final assembly.

Aluminum Bumpers Are Best

My latest car, the Red Devil, has an aluminum front bumper from Auto Metal Direct. I had it painted body color and flushed the bolts on it so it looks a lot like an Endura bumper, but without the cost, hassle or weight penalty. So here's the rundown: they look great, weigh nothing and are reasonably priced—that sound's like a winner in my book.

To flush mount the bolts, I found 3/8 inch aluminum bolts (not easy to find!), welded them on the bumper and ground the bolt heads flat—and that sounds way easier than it is, but it was worth it.

Even more than the looks, I like aluminum front bumpers because the entire bumper weighs 3 pounds!

Carpet Install

How to install a new carpet the right way

Installing a reproduction carpet kit in your Pro Touring car should be a no brainer, right? Well, actually, it can be quite involved because many times there are multiple new things bolted or welded to the floor that the carpet needs to be situated over/around/behind/etc. I used to torment myself with this task until I found an interior shop I really trusted and got the job done quickly. So, if you are like me and you want to try to do this job yourself, I shot photos of my friendly upholstery guru, Jerry Stewart at All American Auto, installing a carpet kit from Year One in the Mule—use his years of knowledge as a guide to help you end up with an interior carpet that looks great and does the job.

Installing a reproduction carpet kit in your Pro Touring car is a must for comfort and appearance. Start by removing the carpet from its shipping box(es) and laying it out to let the creases, bends and folds 'relax' out. Don't start working until the carpet 'lays flat'.

Top Tips for Installing a Carpet Kit

1. Buy a quality carpet kit for your specific vehicle
2. Spend the time to get the front carpet section properly situated in the car before starting any trimming—this is the top mistake: not getting the carpet sitting in the car correctly before starting the trimming
3. Measure, remeasure, mark up (with chalk on carpet side, white wax pencil on back), then remeasure—before cutting!
4. Cut conservatively, refit, remeasure, then cut a little more if needed.

As I showed back on page 154 of this book, I suggest installing sound deadener and heat resistance materials on the floor (this is from Dynamat) before installing your carpet. This will make driving your Pro Touring car much more enjoyable!

The first step is to position the carpet properly in the vehicle. This means putting the front half in the front of the vehicle and the rear section over the back of the front seats/footwell and front of the rear seat area. This is where you'll start the process of getting the carpet to fit cleanly around all the new components you have installed to update your car (rollcage, subframe connectors, late model shifters, etc).

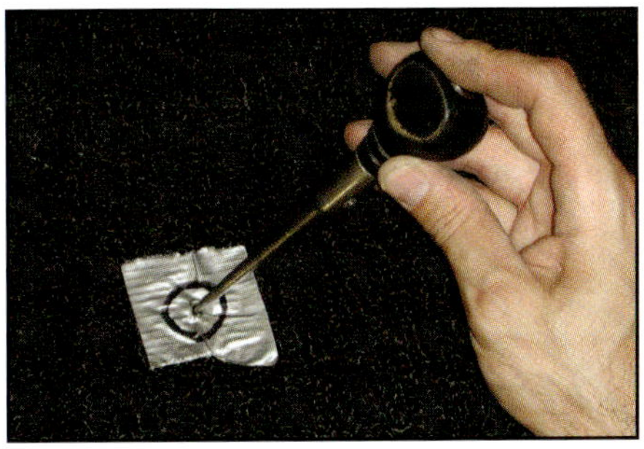

With the carpet properly positioned, Stewart marks the locations of components situated under the center of the carpet, like the parking brake and shifter handle, and 'lock' the carpet in place. After determining the hole location by using a pick tool to feel where the handle(s) need to protrude through the carpet, Stewart uses white chalk to mark the location with an 'X' and draws a shape of what the opening should look like.

Here is the opening for the shifter. Notice how the opening has multiple cuts in it where Stewart was 'sneaking' up on it. He uses very sharp upholstery scissors to make these cuts. When the carpet is being final installed Stewart will cut this into a circle around the shifter mechanism to clean up the opening.

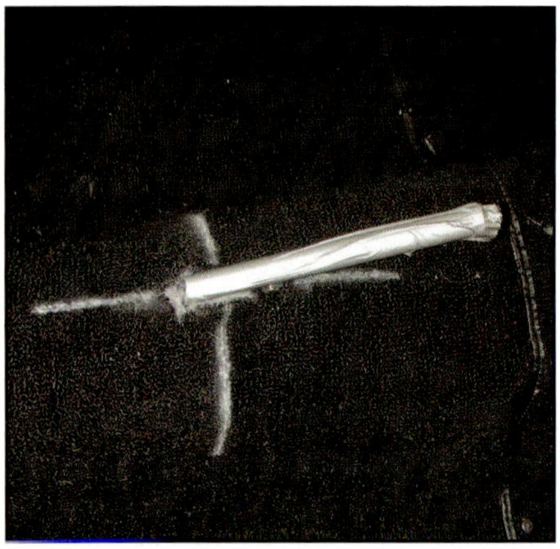

The location of the front piece of carpet is set (the shifter spud hole set that in stone!), so Stewart does the same with the rear piece of carpet by cutting the hole for the parking brake handle. This hole is more long and thin, but it is also 'found' using the pick tool.

With the carpet properly positioned, Stewart marks the locations of components situated around the perimeter of the carpet, like this rollcage downtube, on the backside of the carpet with a white wax pencil. This is in preparation for cutting an opening in the carpet for the component and also cutting an access slit so he can install the carpet around the component.

The carpet section is then removed from the vehicle and flipped over so the wax pencil marks can guide the cuts. As a tip, the access slit is cut longwise because Stewart says he would never be able to get the short section to 'lay flat' where it was supposed to if he cut it into the shortside. EXPERIENCE RULES!

So here is what the access hole and installation slit looked like after Stewart cut them into the Year One carpet. The slit runs back under the rear seat, so the seat frame will help to hold the carpet in position around the rollcage tube.

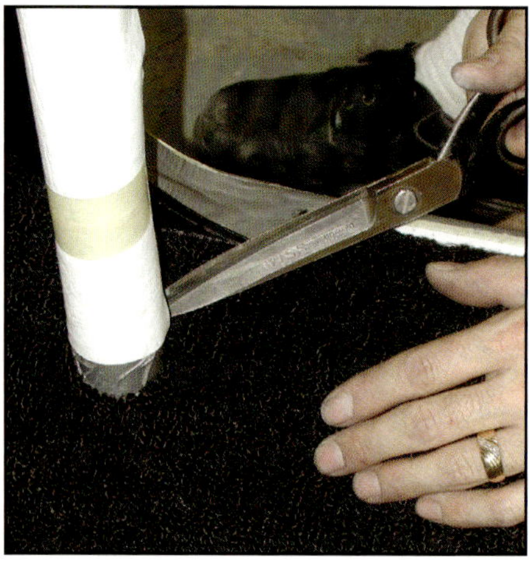

Once the carpet is rough installed around the component, trimming is done on the carpet with it in position around the component.

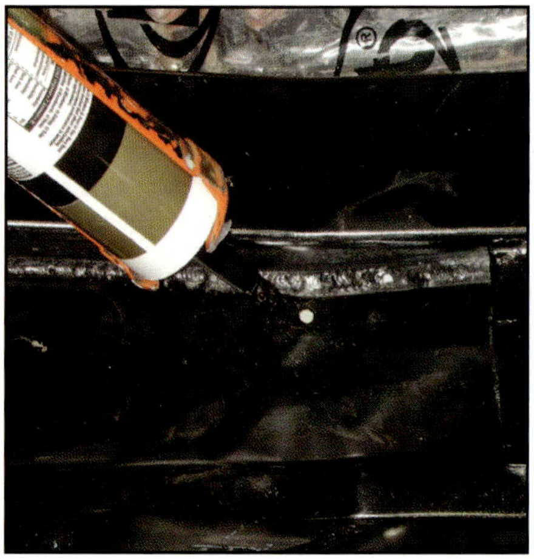

Stewart then removes the carpet and goes to work preparing the floor for the final install of the carpet. The first step of this is to seal any seams or holes with pump-in sealer.

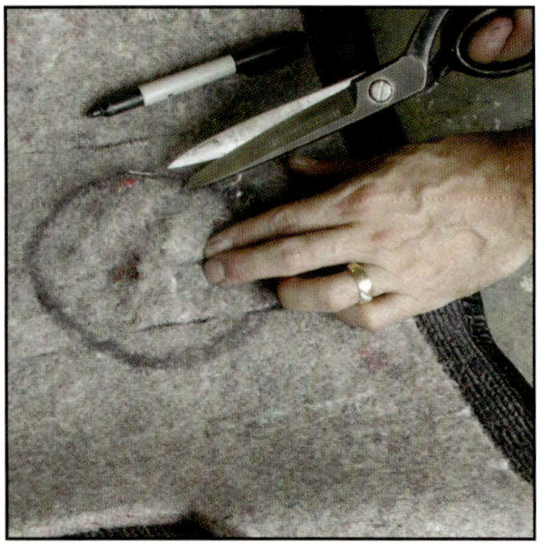

With the carpet laying on the floor face down, Stewart trims the jute padding off the back of the carpet section where the shifter will pass through.

Trimming the jute requires multiple passes with scissors, as you can see here. You just keep trimming away material until you are down to the backside of the carpet. This minimizes the chance of the jute getting stuck in the shifter mechanism.

Speaking of jute, Stewart likes to position it on the footwells in the floor to improve the 'feel' of the carpet. Here, he has cut two sections of jute to fit into the rear footwell that has been bisected by the DSE subframe connector.

Stewart then sprays his super-sticky adhesive onto the back of the jute, letting it get tacky before installing it.

He also sprays adhesive onto the floor at the same time as he coats the jute. This means when the jute is placed on the floor, it isn't moving anywhere—which raises the stakes of the install! Another good reason to have a pro do this—they get pieces like this installed correctly on the first shot.

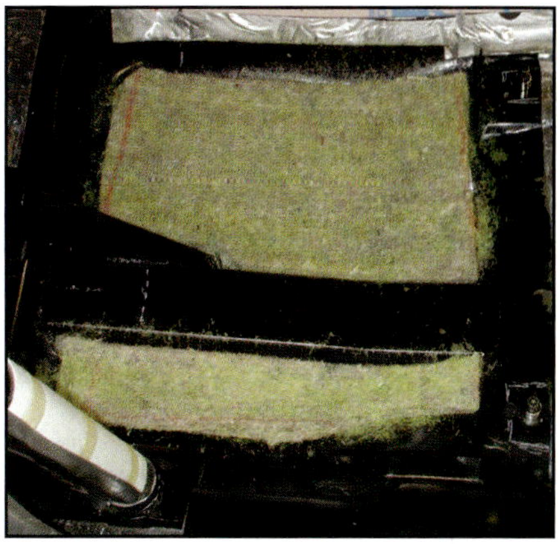

Here is the jute properly installed in the rear footwell. Notice the foam protection around the rollcage tube to minimize the chances of the paint being damaged during the interior work…nice touch!

With all the trimming and jute installation completed, Stewart positions the front carpet section over the shifter and begins to install adhesive to both the carpet and floor—holding the carpet up until the adhesive tack-dries.

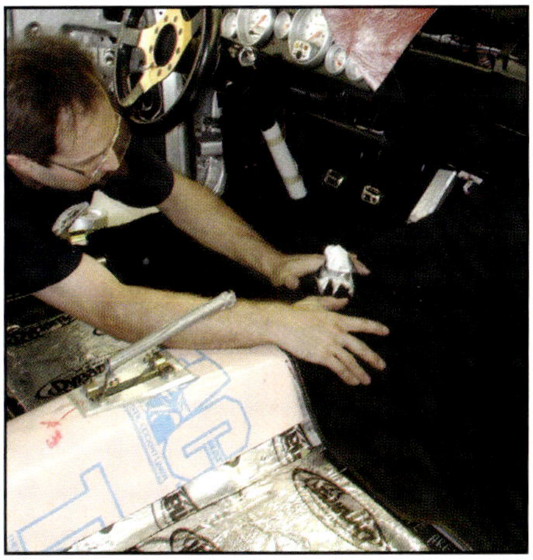

Stewart does the final install of the front carpet section from the center of the car to the outer perimeter of the car. Once the carpet is installed, he works out any irregular surfaces or issues by pulling the carpet or massaging it into position.

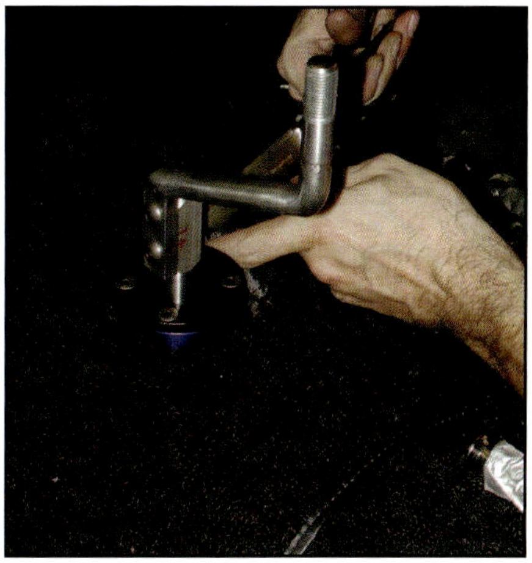

The final 'trim arounds' are done at this time. This means areas like the transmission shifter hole are cleaned up so the carpet fits tightly around the shifter mechanism (even though this will be covered with a shifter boot, Stewart likes it to be tight to the shifter).

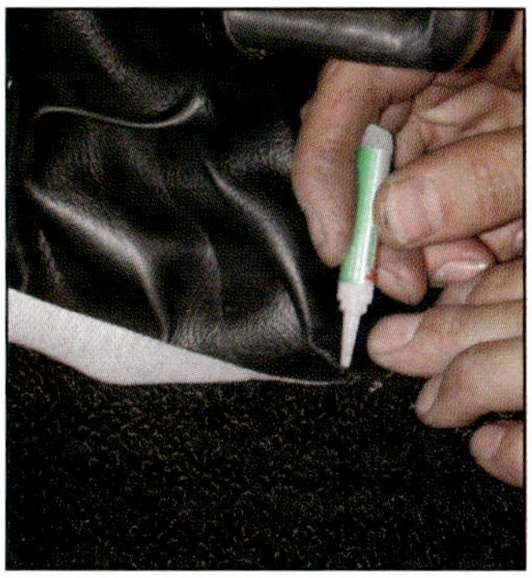

Stewart custom-sewed the shifter boot out of high-grade, black leather. Here, he is marking to finalize the shape.

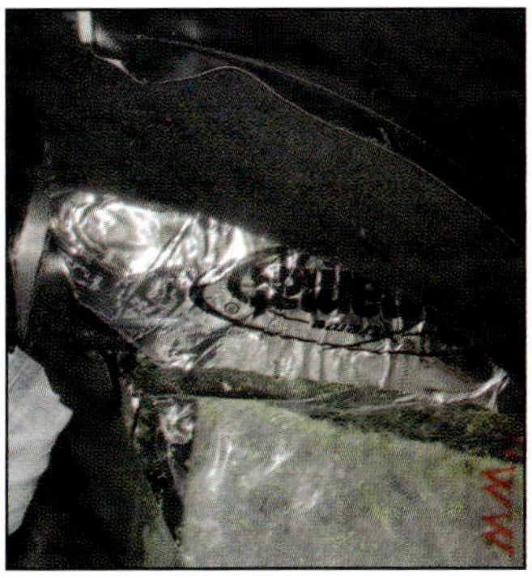

Modified areas like the rear inner wheelhouses (which are widened) require additional massaging to look and work properly. Besides being covered with plenty of Dynamat, Stewart cut a piece of closed cell foam and glued it to the wheelhouse to further mute noise from coming in.

Certain areas, like the passenger side underdash area, require a little help to look their best. Stewart thought there was too much firewall exposed (this is usually covered by the stock HVAC system), so he built a little carpet extension.

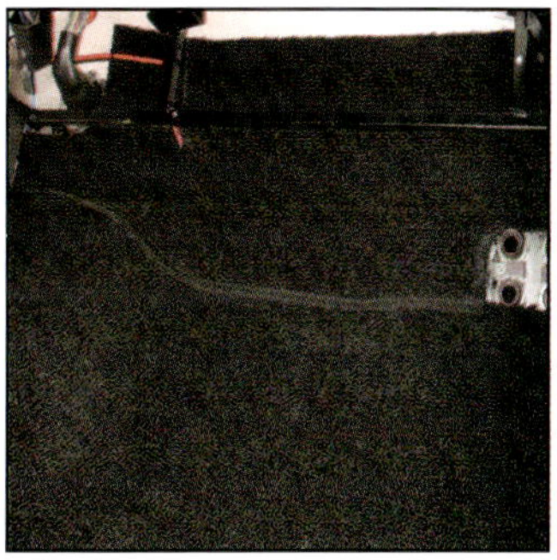

This is the carpet extension installed under the dash. The Vintage Air A/C unit will hide a lot of this, but I had noticed on my previous cars that the carpet seemed to 'stop short'. Notice the A/C hose bulkhead fitting on the right—placed there so I can route the hoses through the passenger side fender, and the F.A.S.T. controller wiring bulkhead on the upper right—which hid the engine harness wiring under the hood.

Tail Light Tuneup

Fixing up the final little details, like the running and tail lights

I do like to run as many original equipment parts as I can, so when there is an opportunity to clean up a part, apply a little paint and polish, then reinstall it, I'm there. With the tail light housings, that is exactly what I do.

Now, as with other rubber parts on my cars, the stock tail light housing seals are usually junk, so I replace them with Year One seals when I reinstall everything. Other than that, this process goes quickly but makes a nice difference in the appearance and function of the tail lights (and running lights).

You'd be amazed at what a little scrubbing with some 3M Scotchbrite pads followed by some painting of the parts can do to the appearance and function of your tail and running lights—here is what I do to make them work great.

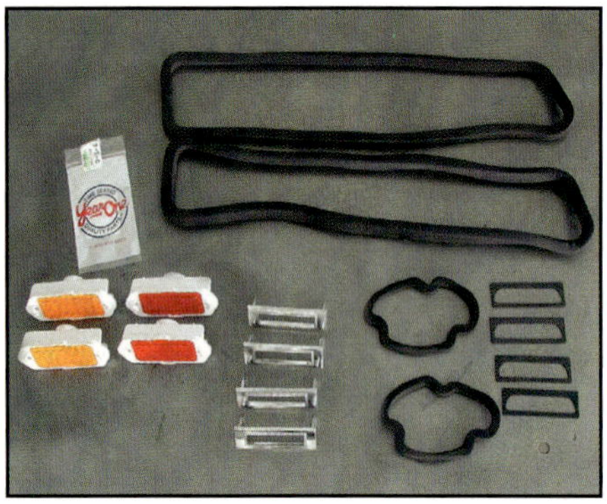

Here are the parts I always order from Year One when doing a Pro Touring car: new tail and running light seals, new side running lights (these seem to always be broken) and chromed housings, and at least one new light socket (the springs in them break after a while). I sometimes order more (fasteners, etc), but this is the 'for sure' list of parts.

The tail light housing on the left shows what you can achieve with just a little elbow grease and some paint. I find the original tail light housings are usually reusable on early musclecars.

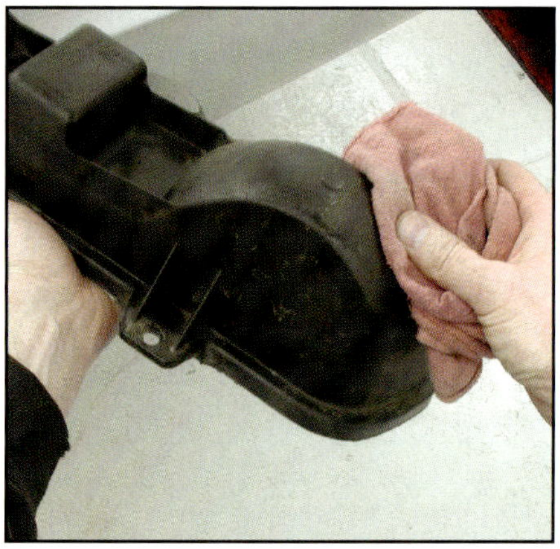

I start by wiping down the housing on the inside and outside (shown) with a solvent soaked rag to remove any dirt, overspray or other gunk.

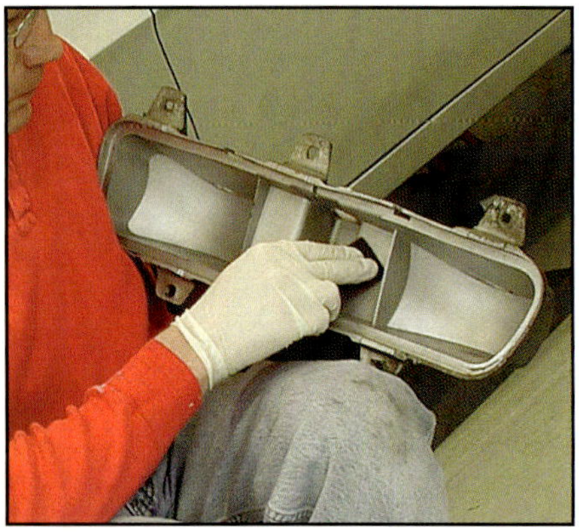

I then rub every surface with a Scotchbrite pad to prep the surface for painting. This usually knocks off any loose paint or removes any irregularities in the surface of the plastic.

Before painting the inside of the housing, I tape over the entire backside of the housing to prevent any overspray (the inside is a bright silver, so the overspray would look especially bad). I just paint on my garage floor, so I lay down some newspaper around the part.

I like to use Eastwood's Reflective Aluminum paint on the inside of the tail and running lights. It covers quickly, looks great and has shown itself to be very durable.

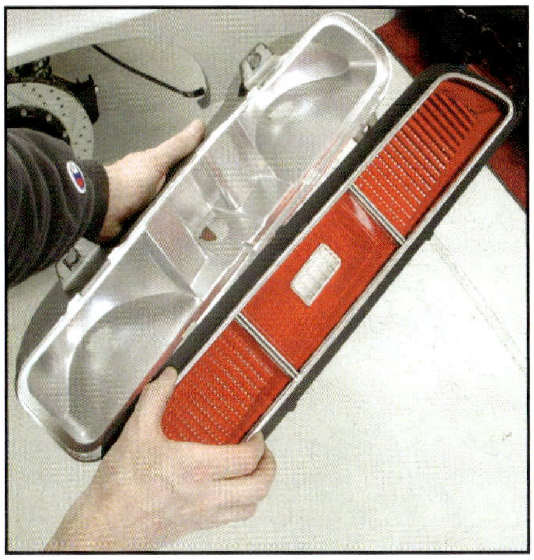

Here is the reconditioned tail light housing with a Year One tail light lens being installed. Looks pretty good, doesn't it?

The tail light housing requires a seal that is sandwiched between the body and housing. I ordered that early in the build process from Year One as it was one of the many components I knew I would need later on.

The installation of the tail light housing is usually as simple as sliding the housing into the body opening and tightening the fasteners (which I am sure you bagged and marked for reinstall at this time, right?).

This is what the Mule Camaro looked like after I completed the install of the tail light housings. While this shot might not look like anything special, I am always impressed with how good a Pro Touring musclecar looks with all the gaps even and parts fitting properly!

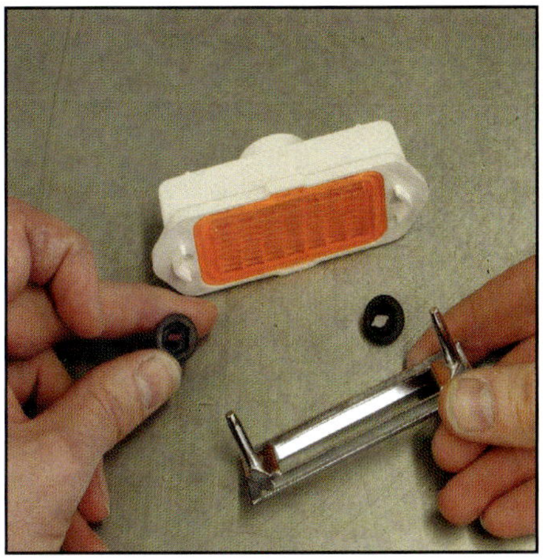

This is part of the reason I buy four new sidelamps—the plastic studs on the frames usually break off trying to remove these self-threading nuts (in left hand). No problem, Year One is ready to support when needed…

The running lights that install in the front fascia are usually in need of some help on any car. They take a lot of abuse from debris hitting them and dirt getting stuck in the light fixture, so I often replace the sockets and lens on them.

The springs on these contactors are often bent or broken. If they are, I just replace the socket with one from Year One as I have not found a way to fix them (as quickly or inexpensively as replacing them).

With a new lens, seal and socket, the running lights are like new and easy to install. These usually install easily into the fascia and require very little finesse to end up looking great in the body opening.

As a tip, I usually use either the stock wiring or wiring from Year One to connect the running light to the front of vehicle wiring harness. That's another advantage of using the Painless wiring harness—these little connectors plug right in! That's nice.

Cool Headlight Install

I admit it. I like to install new parts I think will improve the performance or appearance of my cars. These headlights, from the now defunct Iguana Motorsports, would fall under that heading.

While Iguana is not around any longer, I do think all Pro Touring cars should seek help in the headlight department. I know many folks that just replace the stock headlights with a halogen headlight that fits in the stock headlight bucket. On the Mule, I chose to convert to a High Intensity Discharge (HID) headlight system. Whatever you do, an upgrade is needed to keep up with the speeds now possible with Pro Touring cars.

I have used the Detroit Speed Bright Driver headlights with good success. They install easily and provide excellent light coverage. There are other companies, like Hella, that offer headlight upgrade kits. Whatever product you use, I definitely recommend making an upgrade as ANY Pro Touring musclecar will overrun the stock headlights—and that is just plain unsafe.

As Pro Touring cars get faster and faster, they need more powerful headlights. Converting to a more powerful late model headlight often requires some rework, but it is worth it. This is how I got a HID halogen upgrade kit to fit in the Mule

This is how I packaged these no-longer-available High Intensity Discharge (HID) headlights in the stock '69 Camaro headlight buckets and mounted the required-relays on the side of the headlight bucket.

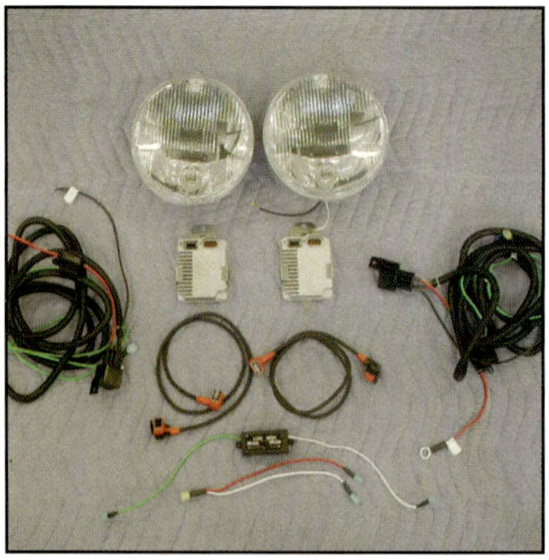

The company that offered these, Iguana Motorsports, is no longer around, but I liked how they provided the headlights, wiring harnesses and ballasts with the kit. I believe these lights were discontinued due to D.O.T. issues.

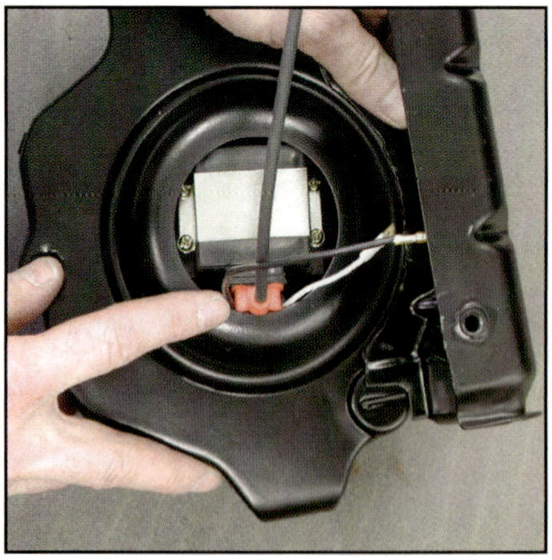

Here was the biggest snafu I ran into installing these HID headlights: the light didn't fit in the stock rear opening of the headlight bucket. I could have just ground away material until it fit, but that would have looked butch, so I came up with a better, but more involved, solution.

Here's my solution—use a large 4 inch diameter hole saw to cut the opening larger. I have the headlight bucket and hole saw on my Bridgeport mill, but this could be done on a low-dollar drill press.

The opening needed to be made off center in the backside of the headlight bucket, as you can see in this shot. Once the hole was made, I dressed the edges with a machinist's file (to eliminate burrs and sharps) and painted the bucket black again.

When you go to High Intensity Discharge (HID) lights, you need to run transformers to step up the voltage to the HID lights.

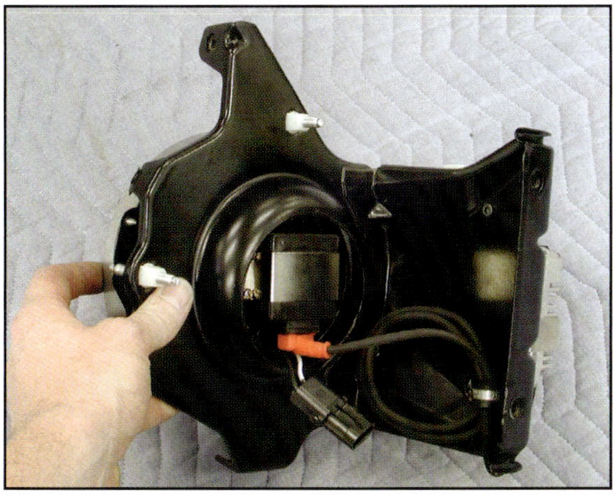

Here is what the headlight assembly looked like just before being installed on the Mule. I think I had about 4 hours in each of the headlights to get them all installed in the car and working! I don't think that is unusual…it just takes time to get everything to look and work like it was meant to be.

Once I got the headlights installed in the car, the wiring needed to be strung through the front end. This could go quickly if I wasn't concerned about how it all looked when done—which is not the case. I ended up reworking the entire front wiring harness to incorporate this system.

One of the technical aspects of this system was that the headlights had HID low beam and a halogen high beam lights. The kit came with this controller that I wired into the vehicle to power the two different lights.

I think the look of the Mule was heavily influenced by the upgraded headlights. They look familiar, but better, which to be honest is what I shoot for with all of my upgrades. They are just enough to make you scratch your head and wonder what is different

The Mule – Beauty Shots

The Payoff

Photos used with Permission by GM

The Mule was a hit on many fronts. A special experience was when I was asked by GM Design to allow them to put the Mule in their studio in 2005 as a reference while the 2010 Camaro body was being sculpted. Here it is being photographed by the GM Design staff—they let me stand in for one photo!

I have built many Pro Touring cars in my life that I thought were impressive, good looking cars, but one car stands out from them all in the minds of the public: the Mule. And I think I understand why. While the Mule was a huge leap forward for Pro Touring cars and for me as a car builder, it has a timeless look. In fact, the Mule looks like a Pro Touring car that could have been built last winter—even though it was built almost 10 years ago! That alone tells me it was way ahead of its time and set a lot of trends in motion that are still resonating with enthusiasts today.

The level of fame the car garnered once completed was something I had not seen with my previous cars. It appeared in a BF Goodrich ad (see sidebar 'BFG to Fame'), was a finalist for the MOTHER's Shine Award at SEMA and was on the cover of Popular Hot Rodding two times (besides appearing in 23 consecutive technical articles showing various aspects of the buildup).

In the days when the Mule was completed, there weren't any Pro Touring competitions like there are today, so I didn't win any Pro Touring track events with it. But we did take it out to many Track Days at local tracks—the best story I can tell you about our track day success was when we went out to a local track during the yearly Ferrari club track day—smoking those guys in a '69 Camaro was hilarious. They were not amused and we didn't stick around too long, but we made an impression!

I continue to build cars and will show you my latest in the next few pages: it's call the Red Devil and won a bunch of awards and competitions in 2010, but the Mule still holds a special place for me. It is definitely a winner.

Shoot the Beauties

I learned this from my friend Will Handzel: when a car is completed, make sure you get a lot of 'beauty' photography before you drive it anywhere. And I mean shots of everything—especially the underside of the car. Find someone you know that is good at photography and have them shoot every shot possible. That means every conceivable angle, light setting, background, underneath, underhood, car-alone, car-with-you shot you can imagine…then dream up some more and shoot those.

You can never have enough beautiful photography—and now video—to document the final product. Whether you just want to show your kids or your grandkids the car you built years ago, or you are lucky enough to get a feature published on the car, having tons of high-resolution, professionally shot photos and video of your car is invaluable. Here is a quick list of some of the setups I always have shot of my cars when they are done:

-Underneath *(Suspension, transmission/exhaust/ engine area, fuel tank, subframe connectors, etc)*

-Underhood *(Engine overall from multiple angles, closeup of 'cool' features)*

-Overalls *(Front, Side, Front ¾, Rear ¾--all shot at ground level, eyeball level, 8 feet off ground, etc)*

-Interior *(Dash, Seats, Rollcage, Details, Trunk, etc)*

-Action *(Car-to-Car, Pan Blur, Dragstrip, Road Course, Autocross, etc)*

-Etc *(Technical Details of everything inside and out)*

BFG to Fame

This only has happened once in my life—the Mule was used as the car in the BF Goodrich tires advertisement for a year of promotion. The whole process was pretty amazing. I was sent drawings of what the ad would look like when completed (inset, below left) and the Mule was transported to an indoor photo studio where it was lovingly photographed for an entire day (see inset, lower right) to end up with the photography to support the ad. The smoke on the rear tires was added in the 'postproduction' work via Photoshop.

I think this only added to the recognition the Mule has had over the years and was really a neat thing to be a part of. As with most experiences like this, you don't realize how special something is until years later looking back—like now!

These gorgeous photos were used as a reference by the fifth-gen Camaro team as they were crafting the shape of the next Camaro.

Obviously, the Mule had been on the streets for a few years when the GM Design photos were taken (as can be seen by the '03/04/05 HOT ROD Power Tour stickers in the back window)—but it still looks amazing in these shots, doesn't it?

I don't vary too much on my interiors, so the Mule doesn't look that different from my other cars. But it is comfortable, makes it easy to drive fast or cruise and still looks sharp after all the years/miles.

The 1000+ hp twin turbo small block has done multiple HOT ROD Power Tours since it was built and still provides insane propulsion for this ride—and it looks great when the hood goes up too.

It looks simple, but to fit a 335/35R18 Michelin Pilot Sport into the fender opening of a '69 Camaro like it is a stock wheel/tire combo was an incredible feat in 2001.

The front suspension was ahead of its time when the car was built and still looks purposeful and track ready.

This is a common look for my Pro Touring cars: exhaust and subframe connectors along with lightweight components, like the aluminum driveshaft, wherever possible.

It still amazes me that my idea for a one-off rear suspension is now considered a must-have item if you are going to build a Pro Touring car. I built it because I wanted a '69 Camaro that handled better—apparently a lot of other people want the same thing too!

The Red Devil

Fast forward to today—literally

Here it is: the baddest 1969 Camaro in the land…the Red Devil. I drove this car to victories in most of the big Pro Touring competitions in 2010.

My latest Pro Touring car is a 1969 Camaro I named the 'Red Devil' because it is painted red and blindingly fast. When it was completed in the spring of 2010, I had a lot of people call it 'just another '69 Camaro Pro Touring car'. Well, it is a '69 Camaro, but it's probably the most refined, brutally quick and fast '69 Camaro in the world.

In case you question that statement, you should know the Red Devil ran up front in many of the big Pro Touring competitions held in 2010: 2nd in the Motor State Challenge, 1st at the LS Fest Shootout and 1st at the Optima Challenge. At each of these events, I was competing against every kind of vehicle on the planet, including late model (C5 and C6), race prepared Corvettes—and the Red Devil still won two of the three events overall! So, people can say it's 'just a '69 Camaro' all they want, but I doubt the guys I beat would say that!

The Red Devil started like all my cars—as a rather simple looking, rust-free, driver '69 Camaro out of the Southwest.

The Future Today

Seeing how I built the Red Devil really shows what the Pro Touring business has become since I built the Mule (back in 2001). Unlike the Mule, the Red Devil was built with a lot of off-the-shelf items: Detroit Speed & Engineering (DSE) front and rear suspensions/steering/subframe connectors, Thomson Automotive 760+ hp 427 ci LS9/7 engine, a lightweight, aluminum hood and bumper and a bottom feed spoiler system from TechAFX. It amazes me how well these parts go together to create a Pro Touring car with stunning performance.

I look forward to the next steps the Pro Touring marketplace takes in the future and definitely plan to be a part of it!

This '69 Camaro housed a factory installed, unmolested small block V8—which I like because it means the car was not 'hacked up' before I got to it.

Yeah, this is what I search high and low for: completely solid sheetmetal underneath. Pretty amazing considering this car is 41 years old! There were dings and dents all over the body, but otherwise this 1969 Camaro was an absolute find.

When this Camaro was built back in 1969, the front subframe/radiator/engine/transmission/etc was loaded into the car from underneath as one unit. I decided to remove all of those things in the same way and test fit the DSE front subframe: it slid right into place.

The Red Devil was prepped and painted by the father & son team of Paul & Joe VanNus with Standox paint products. They did an incredible job smoothing a rough body and getting everything covered with Standox primer so I could do my mockup work on a straight car.

The lightweight aluminum hood from Auto Metal Direct took some fitment rework to sit properly in the opening—not unusual when mating to a car built in the '60s—it works and looks great.

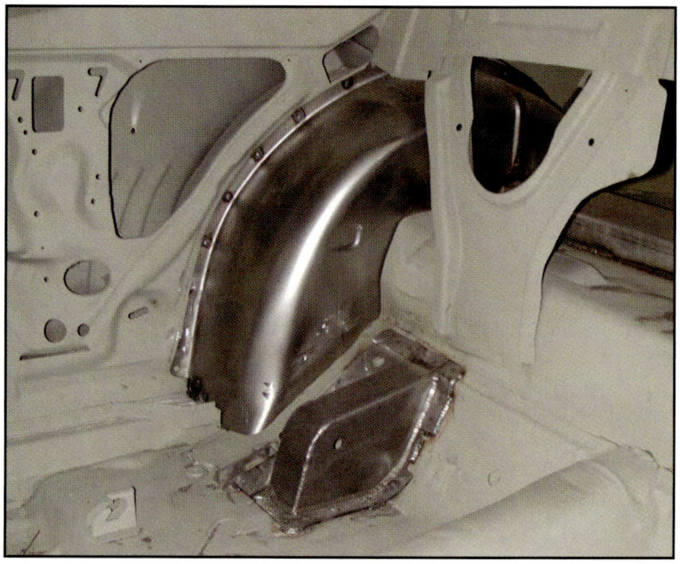

As I was resolving the issues of switching to the LS9/7 engine, I sent the car out to have the DSE Quadralink four-link rear suspension and a set of the DSE Deep Tubs installed in the body.

Here is where I ran into a problem: the engine I originally planned on powering the Red Devil with was a hopped-up 600 hp LS7 from SDPC. The more I thought about this, I realized this wasn't going to be enough to really make a statement—so I decided to go with what I call the LS9/7 engine: a 760 hp 427 ci LS9 killer!

Here is the DSE front subframe/suspension with an LS7 block/LS9 blower & drive (available from SDPC) mockup engine & T56 manual trans to help determine what I needed to modify to make everything work. The big changes were adding more heat exchangers and mounting the Vintage Air A/C compressor.

The LS9/7 engine required positioning the Vintage Air A/C compressor where it hit the passenger side of the DSE front subframe crossmember. To resolve this I built my own A/C mount, then…

I cut the engine mount face off the DSE crossmember to create space for the A/C compressor. I welded in a plate at a lower point and used a Gen I V8 engine mount and adapter to that side of the engine to make it all work.

To mount the Vintage Air A/C compressor to the LS9/7 engine, I built two 'outrigger' brackets from 1018 mild steel 1/8 inch thick strap and two short pieces of ½ inch outside diameter (OD), 0.065 inch wall round mild steel tubing to capture the 'through' bolt.

Once the body was sanded smooth and primed, I brought it back to my shop and mock assembled everything into what I first envisioned the Red Devil to be—

Once I had everything gathered up, the 'real' final package started to come together. As you can tell, I still assemble my cars completely multiple times before they are 'final assembled'—this way I know they will go together the right way in the end.

Even though I know it will come apart again, I act like this is the final assembly process to make sure everything goes together 'right'. One aspect is getting the DSE front subframe properly situated in the car—my friend Nathan Sumner is using a plumb bob off the fenders and measuring to the framerails until everything is 'square'.

The LS9 blower inlet points in a weird direction (up and towards the center) for the first-gen Camaro, so I decided relocate the inlet towards the driver's side. Here I am mocking up where the throttle body will go.

I then cut the throttle body flange at the desired angle to check my fitment. We then removed the intake, made a simple fixture to hold everything in place, cut the intake inlet back much further and master welder Whitey Brown welded a throttle body flange onto the LS9 intake—with 40 feet of aluminum welding rod!

This is what the Red Devil LS9/7 intake throttle body mount looked like after a lot of grinding and finishwork. You can now buy this inlet from Thomson Automotive in Wixom, MI.

Joe Borschke, at Stenod Performance, then built this 4 inch diameter aluminum inlet tube to connect the LS9/7 to the TechAFX Airbox Closeout (for '67-69 Camaros) and AEM air filter (pn 21-3059DK). With that, I was done with the mockup LS9/7 engine and ready for the real powerplant.

Back at Thomson Automotive, Brian Thomson (left) and master calibrator Dave Mikels, were hard at work making the LS9/7 twist the dyno to over 760 hp. This engine is amazing—it just makes obscene power with no worries whatsoever.

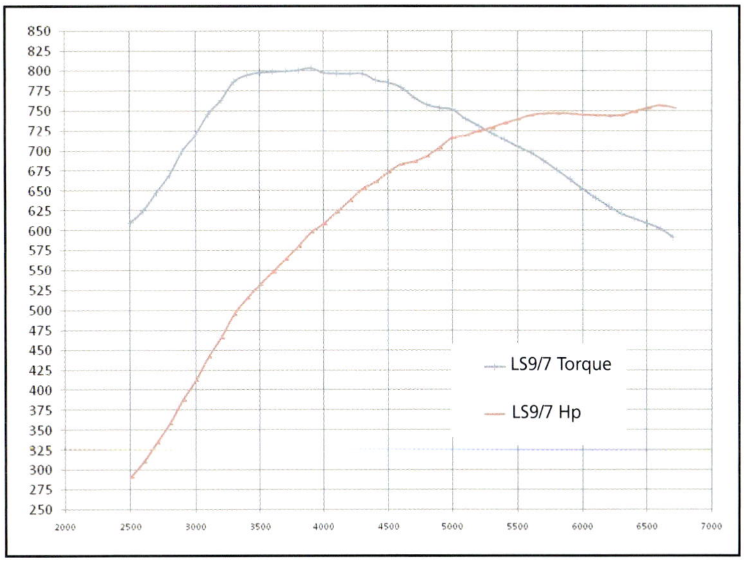

I wanted the top of the blower to be specific to this engine, so I had the supercharger cover modified so it says '7.0L SUPERCHARGED' instead of LS9 SUPERCHARGED'. Thomson Automotive now sells the LS9/7 engine with this cover.

To modify the blower cover, I asked my friend Craig Adams at CADesign to create an insert I could weld into the supercharger lid. He created a 3D computer file of the insert that we had CNC machined out of billet aluminum

This computer 'screen capture' shows the blower with the '7.0L' insert in it.

Adams gave me specs on the opening I needed to machine in the supercharger lid to accept the '7.0L' insert. This is a screen capture of the computer model of that opening. I just machined this into the cover, dropped the insert in place, welded it up and machined the weld flat.

I then put master fabricator Whitey Brown to work packaging up all the heat exchangers (inlet air, A/C condenser, radiator, transmission, power steering) on the core suppport—providing enough cooling for everything was one of the big challenges on this car.

I also had every opening in the core support welded shut to maximize the amount of air going through the heat exchangers. At this point, I disassembled everything and sent it back to the painter for final prep and painting.

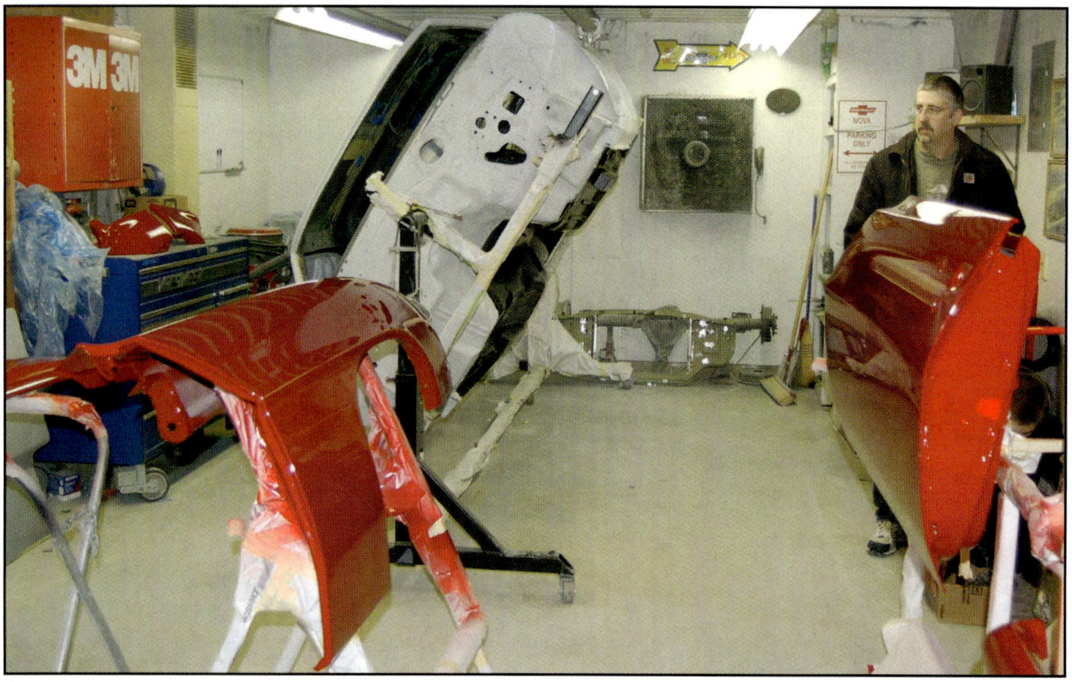

Back at VanNus's paint shop, they had pieces of the Red Devil being prepped/painted/polished just about everywhere. While their place is small, they make excellent use of every square inch—and their prep/paintwork is beautiful!

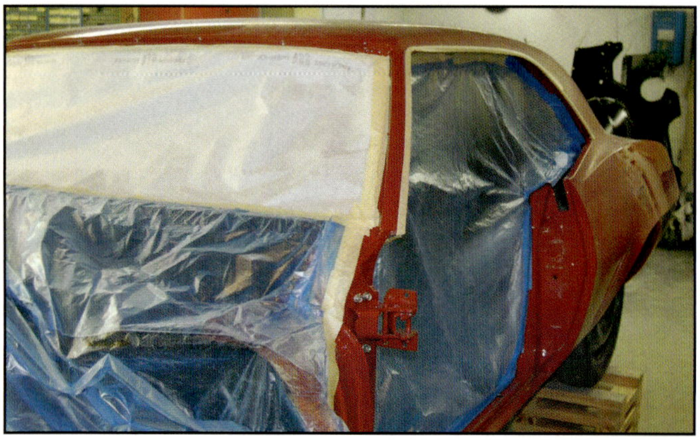

As a customer, this is the type of detail you like to see at a paintshop: The body shell, door jambs and trunk sill are being final painted and everything else has been taped off & bagged. Also, the shop is immaculate and parts are quickly moving towards completion.

As with many of my previous cars, I took all the factory fasteners that are reuseable (not stripped or with mangled heads) and had them chemically stripped, then cadmium plated. I reassembled the Red Devil with a lot of its original fasteners.

The first job once the car was painted and back in my shop was to install my latest innovation: an Antilock Braking System (ABS). I installed a 2006 Z06 Corvette ABS module on the DSE subframe crossmember and plumbed all the hardlines to the wheel cylinders.

To mate the LS7 crank to the T56 manual trans, I needed a special flywheel and clutch: which is available from Lingenfelter Performance. These are expensive, but man are they beautiful...and they bolt right on (which never happens). This sucker has taken everything I've throw at it without a whimper.

Here's me lowering the LS9/7 into the Red Devil for the final time. Notice the engine wiring harness (being built by Ryan Kuhlenbeck) and the LS1 engine mount I used to allow the Vintage Air A/C compressor to mate with the LS9 front accessory drive.

I had to relocate the DSE Delay Wiper motor up 3/8 inch to clear the LS9/7 engine. Also, the plastic tank on the passenger side of the firewall is the vapor separator for the factory 'dry' sump oiling system (the remote oil tank is mounted in the front on the passenger side).

This is the Red Devil as it came together in final assembly. As you can see, I lay multiple, ultra-soft terry cloth towels on the painted and plated surfaces to protect them from damage when doing the final assembly of my cars.

I created a 'bottom feed' spoiler system on my previous car in an effort to improve the cooling capability, aerodynamics and appearance. TechAFX took my idea and created a slick Bottom Feed Air spoiler system that installs in the stock location—it looks and works better than the system I built and installs in a few minutes.

I had planned on doing some testing before competing with the Red Devil, but it was first road ready the night before the Motor State Challenge—so we went there and won the Road Race event. Ha! I later went testing at Milan Dragway (shown here) to further refine the car.

The Red Devil's first day on the road was at Gingerman Raceway in Michigan—competing in the Motor State Challenge! I won the Road Course event and finished 2nd overall. I was happy considering the car had just a few miles on it when we unloaded off the trailer!

The big race of the year for the Pro Touring world is the Optima Shootout in Pahrump, NV. It is a combination event of Road Course, Autocross, SpeedStop Challenge (0-60-0 mph) and Design competitions. I was thoroughly ready for this event and won it Overall—which was my goal when building the Red Devil. Job Done!

Milan Dragway allowed me to do some skidpad, slalom, 60-0 mph and 100-0 mph testing while we were there so I could grab some data from a DL1 data logger. The Red Devil performed amazingly well right out of the box and continues to get better as I get it more dialed in.

I have learned a lot about how to position the wheels & tires in the '69 Camaro wheelwells to get my cars as low as possible. The Red Devil is a result of that work. As a tip, I always leave 5/16 inch space between the tires and sheetmetal to minimize tire rub during hard cornering. With this big of a front tire I do sometimes get slight rubbing under hard cornering.

I like this shot because the Red Devil looks like a '69 Camaro—honoring the original look of the cars is a real goal for me when I build my Pro Touring cars. I like to improve the car at every corner, but keep the overall silhouette 'stock' where I can.

I worked really hard to take weight out of the Red Devil, but am still not happy with its overall mass: it weighed 3450 pounds at the Milan Dragway test. I will continue to remove weight—my goal is to get it under 3300 lbs. I have found removing weight requires constant focus and the weight comes off slowly... but it's worth it.

As you can tell, the Red Devil interior is similar to my previous cars. The new pieces are the electronic throttle pedal and Corvette brake/clutch pedals. The standard fare is the DSE Instrument Panel loaded with Autometer gauges and a Sparco steering wheel.

Yes, the Red Devil only has a 4-point rollcage—Pro Touring rules do not require a cage. A 4-point rollcage with a crossbar provides a good mounting place for the shoulder belts. The Sparco seat and belts hold me in position in front of a stock rear seat that has been modified to fit in the car.

The Red Devil trunk is very useful even though it is heavily modified. Notice the DSE battery mount and Optima battery in the right rear of the trunk (for weight positioning) and the triangle-shaped sheetmetal pieces (arrows) that have been added to the floor to make room for the exhaust tubing.

To make it easier to work on the Red Devil, I didn't have the wiring 'hidden' as much as I have on previous cars. My good friend Ryan Kuhlenbeck wired the car, placing the Engine Control Unit (ECU) and fuseblock on the passenger fenderwell.

GM Design Center Reference Car

How do you top this!?!?

Photos used with Permission by GM

Tom Peters, who is also the Lead Designer on the 5th Gen Camaro (2010 and later), sent me this photo of the Mule in the GM Camaro Design Studio being used as a reference car by the Designers when they were crafting the 5th Gen Camaro shape. In the center is Mark Reuss (now GM North America President), on the left is Al Oppenheiser, Chief Engineer and there are a host of other big shots clustered around the Mule. Behind the people on the right is a full-size clay being sculpted of the 5th Gen Camaro for their review. This was an incredible time and a real honor.

From Here Forward!

To many enthusiasts, the Mule represents the beginning of the modern Pro Touring industry--I agree with that. While I have built many all around performance musclecars prior to it, the Mule was the car that really took that big step forward: great suspension, insane power, killer looks and the ability to show it all on the track and street. I look forward to the future of Pro Touring and will do my best to be part of the crowd shaping it for years to come. I will see you at the track!   ~Mark Stielow